Tissue Engineering and Regenerative Nanomedicine

Tissue Engineering and Regenerative Nanomedicine

Special Issue Editors

J. Miguel Oliveira
Rui L. Reis

MDPI • Basel • Beijing • Wuhan • Barcelona • Belgrade

MDPI

Special Issue Editors

J. Miguel Oliveira
I3Bs—Research Institute on
Biomaterials, Biodegradables
and Biomimetics
Portugal

Rui L. Reis
I3Bs—Research Institute on
Biomaterials, Biodegradables
and Biomimetics
Portugal

Editorial Office
MDPI
St. Alban-Anlage 66
4052 Basel, Switzerland

This is a reprint of articles from the Special Issue published online in the open access journal *Nanomaterials* (ISSN 2079-4991) from 2017 to 2018 (available at: https://www.mdpi.com/journal/nanomaterials/special_issues/Tissue_Regenerative_Nano).

For citation purposes, cite each article independently as indicated on the article page online and as indicated below:

LastName, A.A.; LastName, B.B.; LastName, C.C. Article Title. *Journal Name* **Year**, *Article Number*, Page Range.

ISBN 978-3-03921-656-7 (Pbk)
ISBN 978-3-03921-657-4 (PDF)

Image courtesy of J. Miguel Oliveira and Rui L. Reis.

Contents

About the Special Issue Editors

J. Miguel Oliveira, professor (BSc, PhD), is a biochemist and principal researcher ("Investigador FCT 2012 and Investigador FCT 2015") at the Portuguese (PT) Government Associate Laboratory ICVS/3B's (http://www.3bs.uminho.pt/users/migueloliveira).   He is the vice president of I3Bs—Institute 3B's (Univ. Minho); the director of pre-clinical research at the FIFA MEDICAL CENTER, Estádio do Dragão, Porto, PT, since February 2013; and the pro-director of the 3B's Research Group, Univ. Minho, PT. Currently, he is a lecturer in the doctoral program of tissue engineering, regenerative medicine, and stem cells (TERM&SC) at UMinho, PT (since December 2013). He is also an invited lecturer at the Faculty of Medicine, U. Porto (since September 2013) and Department of Polymer Engineering, UM, PT (2009—present). Over the years, he has focused his work on the field of biomaterials for tissue engineering, nanomedicine, stem cells, and cell/drug delivery. More recently, he set up a new research line within the ICVS/3B's on 3D in vitro models for cancer research. He has published more than 260 scientific contributions in scientific journals, of which four review papers were produced under invitation. He holds 16 approved patents and has published five books, two Special Issues in scientific journals, and more than 70 book chapters in books with international circulation. He has participated in more than 250 presentations in national/international conferences and has been invited and/or served as a keynote speaker in more than 40 plenary sessions. He has an h-index of 35 and an i10 of 84 and has been cited more than 5000 times. He has been awarded several prizes, including the prestigious 2015 Jean Leray Award from the European Society for Biomaterials for Young Scientists for Outstanding Contributions within the field of Biomaterials. He is very active in the elaboration and scientific coordination of several PT and international funded projects. In addition, he is a member of the advisory/editorial board of the *Journal of Bio-Design and Manufacturing, Journal of Materials Science: Materials in Medicine, International Journal of Tissue Engineering, Journal ISRN Biomaterials, The Journal of Experimental Orthopaedics,* and *Recent Patents on Corrosion Science* and has been a referee in more than 40 international journals.]

Rui L. Reis, professor (PhD, DSc, Hon. Causa MD), is a Portuguese researcher who has been involved in biomaterials research since 1990. His main area of research is the development of biomaterials from natural origin polymers (starch, chitin, chitosan, casein, soy, algae based materials, silk fibroin, gellan gum, carragenan, hyaluronic acid, ulvan, xanthan, marine collagen, etc.) that, in many cases, his group originally proposed for a range of biomedical applications, including different medical devices, bone replacement and fixation, drug delivery carriers, partially degradable bone cements, and tissue engineering scaffolding for a range of different tissues (http://www.3bs.uminho.pt/users/rgreis). He is an FBSE, FTERM, and NAE member. He is also the vice-president of research and development at University of Minho, Portugal, and director of the 3B's Research Group and of the ICVS/3B's Associate Laboratory of UMinho. He is the CEO of the European Institute of Excellence on Tissue Engineering and Regenerative Medicine (TERM). He was also the head of research and development for the Holding Corticeira Amorim SGPS. He co-founded different start-up companies originating from the research and activities of the 3B's Research Group (Stemmatters and HydrUStent). He has been the co-coordinator of four major European Union (EU) research projects, funded under FP6 and FP7 of the European Commission. He is also responsible for several other projects funded by Portuguese, European, and American biomaterial and polymeric industries and for a range of bi-lateral concerted actions. At only 46 years of age, he was awarded an ERC AdG (European Research Council Advanced Grant)—the most prestigious grant available for European researchers in all of Europe—for 2.35 million euros for his project ComplexiTE. The 3B's Research Group has two additional ERC Consolidator Grants (CoG) running with a budget of around 2 million euros each. Under HORIZON 2020, Rui L. Reis presently coordinates the ERA Chairs FoReCast grant (2.5 million euros for 3B's-UMinho), as well as two TWINNING projects Gene2Skin and Chem2Nature (with budgets of 1 million euros each). He is also the scientific coordinator of the recently approved 15 million euro, EC-funded (6 million euros of UMinho budget) TEAMING proposal, "The Discoveries Center for Regenerative and Precision Medicine" (with University College London (UCL), UPorto, UAveiro, ULisboa, and UNova Lisboa). This project is also supported by FCT and by three regional authorities in Portugal (Northern Portugal Coordination Regional Authority (CCDR-N), CCDR-C, and CCDR-LVT) and is supposed to have a total budget of around 100 million euros in the coming 7 years in order to create a new multi-campi Center of Excellence (CoE) in Portugal, with support from UCL in the UK. He also coordinates a large international PhD programme funded by FCT on TERM and stem cells, as well as two structural programmes for hiring highly qualified researchers (around 7.8 million euros) and project with a budget of 850,000 euros to support the abovementioned TEAMING proposal, all three of which are funded by the CCDR-N. As a result of his academic activities, Rui L. Reis has been awarded several prizes, including the ESAFORM 2001 Scientific Prize, Jean LeRay Award in 2002, Stimulus to Excellence Award in 2004, Pfizer Award for Clinical Research in 2007, START Innovation award in 2007, etc. In addition, he was or is a member of several editorial boards of journals, acts as referee of numerous (more than 120) scientific journals, and has been a presenting author, member of scientific committees and organizing committees, referee, chairman, discussion leader in Gordon Research Conferences, and invited lecturer in many conferences worldwide. Rui L. Reis has thus far produced 1023 publications listed in ISI Web of Knowledge, 902 publications listed in Scopus, and 1515 listed on Google Scholar, including around 816 ISI-listed articles published in scientific journals as a referee, with around 70 of those being review papers or editorials. He holds 35 national and international patents (several other applications are ongoing, and one of the awarded patents was selected as one of 15 finalists for the European Inventor Award of 2013). He has also authored 7 books, 6 special issues in scientific journals, around 230 book chapters in books with international circulation and in international encyclopedias, and more than 1800 communications in conferences, almost all of them in international meetings, including around 215 plenary or invited talks. He has presented around 165 invited lectures in other universities or research institutes. Researcher ID (A-8938-2008); Scopus Author ID 56861715700; ORCID 0000-0002-4295-6129.]

Preface to "Tissue Engineering and Regenerative Nanomedicine"

The convergence of regenerative principles in nanomedicine and tissue engineering promises to create new avenues in regenerative medicine. The great advantages of using nanomaterials for controlled drug delivery, surface functionalization, scaffolding, and sensing and imaging applications are now being recognized. Emerging engineering strategies are investigating the potential of synthesizing nanosystems with different compositions and architectures. Multi-modal and multi-scale materials are now being designed by means of incorporating different stimuli-responsive functionalities, peptides, antibodies, and imaging probes targeting specific cells/cells sheets, tissues, and organs, and such advancements offer the possibility of tracking results in real-time. These methods are opening up new treatment possibilities for tackling several aging-related disorders.

J. Miguel Oliveira, Rui L. Reis
Special Issue Editors

nanomaterials

MDPI

Review

Protein-Based Fiber Materials in Medicine: A Review

Kelsey G. DeFrates [1,2,†], Robert Moore [1,†], Julia Borgesi [2,†], Guowei Lin [1,†], Thomas Mulderig [3,†], Vince Beachley [2] and Xiao Hu [1,2,4,*]

[1] Department of Physics and Astronomy, Rowan University, Glassboro, NJ 08028, USA; defratesk6@students.rowan.edu (K.G.D.); moorer1@students.rowan.edu (R.M.); ling8@students.rowan.edu (G.L.)

[2] Department of Biomedical Engineering, Rowan University, Glassboro, NJ 08028, USA; borgesij5@students.rowan.edu (J.B.); beachley@rowan.edu (V.B.)

[3] Department of Mechanical Engineering, Rowan University, Glassboro, NJ 08028, USA; mulderigt9@students.rowan.edu

[4] Department of Molecular and Cellular Biosciences, Rowan University, Glassboro, NJ 08028, USA

[*] Correspondence: hu@rowan.edu; Tel.: +1-856-256-4860; Fax: +1-856-256-4478

[†] These authors contributed equally to this work.

Received: 7 May 2018; Accepted: 20 June 2018; Published: 22 June 2018

check for updates

Abstract: Fibrous materials have garnered much interest in the field of biomedical engineering due to their high surface-area-to-volume ratio, porosity, and tunability. Specifically, in the field of tissue engineering, fiber meshes have been used to create biomimetic nanostructures that allow for cell attachment, migration, and proliferation, to promote tissue regeneration and wound healing, as well as controllable drug delivery. In addition to the properties of conventional, synthetic polymer fibers, fibers made from natural polymers, such as proteins, can exhibit enhanced biocompatibility, bioactivity, and biodegradability. Of these proteins, keratin, collagen, silk, elastin, zein, and soy are some the most common used in fiber fabrication. The specific capabilities of these materials have been shown to vary based on their physical properties, as well as their fabrication method. To date, such fabrication methods include electrospinning, wet/dry jet spinning, dry spinning, centrifugal spinning, solution blowing, self-assembly, phase separation, and drawing. This review serves to provide a basic knowledge of these commonly utilized proteins and methods, as well as the fabricated fibers' applications in biomedical research.

Keywords: protein; nanofibers; biomaterials fabrication; medicine; tissue engineering; wound healing; drug delivery

---

1. Introduction

Fibrous materials, so often used in industrial applications and the textile industry, have now migrated into biomedical research. To date, polymer-based fibers with diameters on the micro- or nanoscale have been explored in drug delivery [1–4], wound healing [5–7], tissue engineering [8–10], and biosensor technologies [11–13] due to their high surface-area-to-volume ratio, mechanical strength, porosity, potential for surface modification, and tunability [13–15]. Equally important in biomaterials engineering, however, is the need for materials to be both biocompatible and biodegradable. Therefore, to maximize these properties, natural polymer-based fibers made from proteins have begun to be developed [16–19].

The appeal of protein-based fibers for biomedical applications stems from the fact that many proteolytic enzymes capable of degrading commonly used natural polymers are already present in the body. In the case of protein-based biomaterials, degradation of these materials leads to the production of amino acids that pose no risk of toxicity and can be reabsorbed by the body [20,21]. In the field of tissue engineering, this avoidance of toxic byproducts is of particular importance since materials must degrade and be replaced by native tissue to achieve complete regeneration [22]. In the field of drug delivery and nanomedicine, protein-based nanofibers may have the ability to store pharmaceutical products and biological molecules without threatening their bioactivity [23–25]. Their controllable degradation through crosslinking or post-fabrication modifications has also been shown to allow for the controllable release of drugs, with no added toxicity from material byproducts during fabrication [26,27]. Blending of proteins with other natural and synthetic polymers can also allow for the development of versatile materials with modifiable degradation and physical properties [28,29]. The biodegradable nature of protein-based fiber materials also supports the efforts of green and sustainable engineering. Such applications reduce the dependence on petroleum-based polymers avoiding the pollution issues caused by the disposal of these materials and their byproducts [30]. Additionally, many proteins, such as silk, soy, and corn zein, are very abundant and easy to isolate [31].

The incorporation of natural polymers in biomaterials has also been shown to enhance cell attachment due to the presence of native cell attachment motifs [25,32]. Thus, the use of protein-based fibers in tissue engineering and nanomedicine has both medical and commercial appeal. Despite these advantages, however, standardization of the mechanical and physical properties of protein-based fibers remains challenging. Such materials have been shown to vary depending on the method of fiber production, the fiber diameter, and the composition of the fiber [33–35]. In order to illustrate the appeal of protein-based fibers and regulate their use, this review serves to provide a basic knowledge of the commonly used materials and methods for the fabrication of protein-based fibers and their corresponding use in tissue engineering, wound healing, and drug delivery. Popular proteins, such as keratin, collagen, silk, elastin, zein, and soy, are given particular attention, as well as current fabrication methods, including electrospinning, wet/dry jet spinning, dry spinning, centrifugal spinning, solution blowing, self-assembly, phase separation, and drawing.

2. Protein Materials

Some of the specific types of proteins that will be discussed include keratin, collagen, silk, elastin, zein, and soybean (Figure 1). These proteins are some of the most common protein polymers used for the fabrication of fibers for biomedical application.

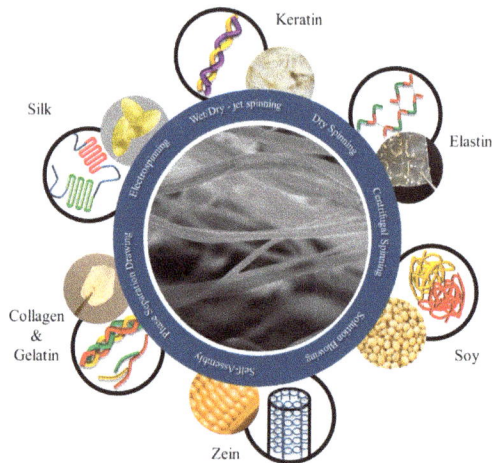

Figure 1. Protein-based biomaterials can be made from a variety of sources. The origin and structures of some of the most commonly used proteins are shown. These include collagen or gelatin, silk, keratin, elastin, soy, and corn zein. These proteins can then be processed into fibers with unique physical properties through a variety of methods.

2.1. Elastin

Elastin is a naturally-occurring protein found in the extracellular matrix (ECM) that maintains the elasticity of connective tissue in the human body. Tropoelastin, the 72 kDa precursor to elastin, is first synthesized by cells in the rough endoplasmic reticulum and consists of alternating hydrophobic and hydrophilic domains. The hydrophilic domains contain lysine residues interspersed by alanine residues, and this arrangement allows for tetrafunctional crosslinking of tropoelastin molecules by the lysine oxidase enzyme. Crosslinking of tropoelastin molecules is further strengthened by self-assembly of the hydrophobic domains that consist of repeating motifs of non-polar residues of glycine, valine, and proline [36]. Complete assembly of elastin molecules occurs outside of cells due to the protein's large size. Tropoelastin molecules are believed to align and crosslink after interacting with extracellular microfibrils near the cell surface. While these microfibrils provide an integral framework for elastin assembly, as elastin-rich tissue forms, these microfibrils become detectable only on the periphery of the protein structures [37]. The elastin-based fibers are then arranged in a variety of structures depending on the tissue's location. For example, in ligaments, elastin fibers are arranged in parallel-oriented structures, but can be found in a honeycomb-like pattern in cartilage [36].

In biomaterials engineering, elastin is used to describe a variety of elastic proteins and peptides rather than one single molecule. It can be used in a variety of forms that include, most commonly, soluble elastin [18,38,39], recombinant tropoelastin [40,41], and synthetic elastin-like peptides that may also be hybrids with other proteins, such as silk [42–45]. The inclusion of elastin in nanofiber scaffolds has been shown to increase fiber elasticity and provide for better cell attachment [18]. Due to elastin's elasticity and resilience, it has found special purpose in the development of vascular grafts, since fibrous scaffolds made from the protein have been shown to closely match the compliance of natural arteries [46–48].

2.2. Collagen and Gelatin

Collagen is a fibrous protein that serves as the main component of the ECM. While the majority of collagen found in body can be classified as type I, II, or III, as many as 29 different types of collagen have been identified. All collagen exhibits a repeating X-Y-Gly amino acid sequence, where glycine is always present as the third residue. X and Y can denote any amino acids, with proline and hydroxyproline being the most common ones. The glycine residue allows for a stable secondary structure formation of collagen, which consists of three strands coiled around each other to form a triple helix. These triple helices can then arrange into different quaternary structures depending on the type of collagen. In fibrillar types, such as types I–III, V, and XI, the coiled coils are crosslinked by the lysine oxidase enzyme to form fibrils that then aggregate to form fibers [49]. The spaces between collagen crosslinking domains measures 67 nm and these gaps give collagen its striated appearance [50].

Due to its natural abundance, Type I collagen is the most common class of collagen used in biomaterials development [49]. The production of collagen fibers has allowed for the generation of biomimetic tissue engineering scaffolds that closely resemble the natural ECM. Therefore, fiber meshes have been used for bone [51–53], cartilage [54], vasculature [55,56], ligament [57,58], skin [59,60], muscle [61], and nerve [62,63] regeneration. These materials allow for cell attachment, penetration, and proliferation due to collagen's ability to interact with cell surface receptors, such as the $\alpha 2\beta 1$, $\alpha 1\beta 1$, $\alpha 10\beta 1$, and $\alpha 11\beta 1$ integrins [64]. The mechanical properties of collagen-based fiber meshes can also be easily modified by chemical or physical crosslinking, although this has been shown to affect their biocompatibility [65].

Due to its complicated hierarchical structure, collagen fibrils can be difficult to extract and isolate. However, the coiled coil can be easily broken down through hydrolysis to produce three polypeptide strands, known as gelatin [66]. These strands can be further degraded into shorter amino acid sequences by matrix metalloproteinases, making gelatin a biodegradable material with low immunogenicity. Due to the presence of alkaline and acidic amino acid residues, gelatin is also amphoteric and can form a thermally reversible network in water [67]. Like collagen, the mechanical properties of gelatin-based materials can be further modified through chemical or physical crosslinking, which is often necessary due to the instability of the natural biopolymer in water at body temperature [68]. Crosslinked gelatin-based materials and fibers made by dissolving gelatin in polar solvents to prevent aggregation [69] have been shown to promote ocular [70,71], bone [72,73], cardiovascular [74,75], nerve [76,77], and skin [78] regeneration.

2.3. Silk

Silk is a natural biopolymer produced by insects, spiders, and worms that consists of two main proteins. Silk sericin, the sticky protein found on the outside of silk strands, makes up 15–35% of silk cocoons and must be removed through a degumming process to extract the more versatile silk fibroin protein [79]. The particular amino acid sequence of silk fibroin can vary depending on its species, but is predominately composed of hydrophobic blocks composed of glycine, alanine, and serine residues and hydrophilic blocks consisting of charged amino acids. The hydrophobic blocks allow for the formation of β-sheets within the protein, giving silk high tensile strength, while the hydrophilic blocks give silk fibroin its elasticity [80]. Variations in specific sequences account for differences in the secondary structure of silk, which, in turn, affects its mechanical properties, thermal stability, chemical characteristics, and solubility [79,81].

Silk fibroin obtained from the *Bombyx mori* silkworm is one of the most commonly used biomaterials due to its availability and low cost [29,79]. It has been shown to exhibit excellent biocompatibility, bioactivity, biodegradability, tunability, mechanical stability, and low immunogenicity, allowing silk-based fibers to be used to create tissue engineering scaffolds that allow for bone [82–85], cartilage [86], heart valve [87], and nerve [88] regeneration. The oxygen and water vapor permeability of silk also encourages its use in wound healing [25,89]. The mechanical properties and stability of silk-based biomaterials can also be modified through methanol treatments that increase β-sheet crystallinity and strength [29,90].

In addition to silk produced from worms, dragline silk produced by the *Nephila clavipes* spider has found use in biomaterials development. Like silkworm silk, it has shown low immunogenicity, high tensile strength, and biodegradibility. Recent studies have outlined the dragline silk's ability to promote cell adhesion, migration, and proliferation of dental pulp stem cells [91] and cardiomyocytes [92], showing its promise as a component in tissue engineering scaffolds.

2.4. Keratin

Keratin is an insoluble structural protein that makes up the bulk of the adnexa of the epidermis, including hair, horns, and fingernails. The protein can be further characterized a soft or hard keratin depending on its amino acid sequence. Both soft and hard keratin, however, have similar secondary structures that consist of two chains, each containing a central alpha-helical domain. These chains are designated as type I and II and interact to form heterodimers that polymerize to form filaments [93]. Some forms of keratin, like that found in hair, have a high content of cysteine residues that interact through disulfide bonding, enhancing the mechanical strength of the protein [94]. Extraction of keratin requires disruption of these disulfide bonds. This can be accomplished through an oxidation of the protein [95].

The presence of cell binding motifs on keratin, as well as its ability to self-assemble, make it an ideal natural polymer to be used in the creation of biomaterials for tissue regeneration [93]. However, because keratin is known to exhibit poor mechanical stability, it is often combined with other natural or synthetic polymers to create composite fibers [96,97]. Such composites have been used for the skin [98], cartilage [99], and bone [97,100] tissue regeneration.

2.5. Zein

Zein is the major storage protein in corn and is a member of the prolamin group of proteins. Its structure and solubility are dictated by its amino acid sequence, which primarily consists of non-polar, uncharged residues, such as glutamine, leucine, proline, and alanine. In addition to its biodegradability and biocompatibility, recent studies suggest that corn zein can exhibit anti-oxidative and antimicrobial properties [101,102]. These properties have led to zein's expanded use in biomedical engineering.

Studies have shown that the corn protein is compatible with human umbilical vein endothelial cells, human hepatocytes, and mice fibroblasts [103]. Neat zein nanofibers have been shown to exhibit low mechanical strength and stability, and the high hydrophobicity of the protein may also prevent cell attachment [104–106]. Therefore, it is often necessary to incorporate additional synthetic or natural polymers and chemical crosslinking to create successful tissue engineering scaffolds. Studies suggest that these composites may promote successful tissue regeneration when used as a scaffold [107,108]. While it may be counter to cell attachment, the hydrophobicity of corn zein does enhance its capabilities as a drug delivery vehicle since it is more resistant to hydrolysis, allowing for longer, more sustained release of pharmaceuticals [109,110].

2.6. Soybean Protein

Soybean protein is a globular protein composed of two main subunits referred to as conglycinin 7S and glycinin 11S. Both subunits contain regions of non-polar amino acids, such as alanine, valine, and leucine; basic amino acids, including lysine and arginine; and non-charged polar residues, like cysteine and glycine. The globular structure of soybean protein makes it resistant to hydrolysis and incredibly stable, leading to its long shelf-life [111]. For biomaterials engineering, the protein is of particular interest due its abundance of functional groups that allow for surface modification and blending with other polymers [112].

Soy protein is biodegradable and can be obtained from abundant renewable resource. In recent years, soybean products, such as soybean whole fat (SF), soy protein concentrate (SPC), and soy protein isolate (SPI), have become alternatives to petroleum polymers due to their abundance and adhesive properties [113]. Compared to other plant protein-based membranes, SPI-based materials are clearer, smoother, more flexible, and have impressive gas barrier properties compared to lipid and polysaccharide formulations.

Although the solubility of soybean protein is relatively low in acidic solutions, solvents with higher pH above 4.8 have been used to process soybean, allowing for fiber fabrication [114]. Due to the presence of ECM-mimetic peptides within the protein, such fibers have seen great success as tissue engineering scaffolds [111]. Some of the most common applications of soy protein materials include skin regeneration and wound healing [115–117].

3. Fabrication Methods

There are numerous ways of fabricating protein-based fiber materials. Table 1 lists fabrication methods along with controlled parameters that affect fiber properties.

Table 1. Parameters of fabrication methods affecting fiber properties [118–130].

Fiber Fabrication Method	Parameters to Control Fiber Formation		
	Solution	Process	Environment
Electrospinning	- Polymer concentration - Viscosity - Conductivity - Solvent evaporation rate - Molecular weight	- Flow rate - Applied voltage - Tip to collector distance - Collector types	- Temperature - Humidity
Wet-/Dry-Jet Spinning	- Polymer concentration - Viscosity - Molecular weight	- Coagulation medium - Coagulation bath concentration - Post-drawing	- Temperature - Humidity
Dry Spinning	- Polymer concentration - Molecular weight	- Post-drawing - Take up speed	- Temperature - Humidity

Table 1. *Cont.*

Fiber Fabrication Method	Parameters to Control Fiber Formation		
	Solution	Process	Environment
Centrifugal Spinning	- Viscosity - Surface tension - Molecular structure - Molecular weight - Polymer concentration - Solvent structure or evaporation rate - Additive	- Rotational speed - Head diameter - Nozzle diameter - Distance from nozzle to collector	- Temperature - Humidity
Solution Blowing	- Polymer type - Concentration - Solvent evaporation rate - Molecular weight	- Injection rate - Gas flow pressure - Distance from nozzle to collector	- Temperature - Humidity

3.1. Electrospinning

Electrospinning is a nanofiber fabrication method that consists of three main components: a polymer solution within a metal tipped syringe, an applied high voltage, and a grounded collector [131–137]. Figure 2 shows two common electrospinning systems (vertical and horizontal electrospinning systems) utilized in current research. In the vertical system, gravity is an important parameter for controlling fiber formation, while horizontal spinning system relies mainly on the electrical force between the spinning device and the collector.

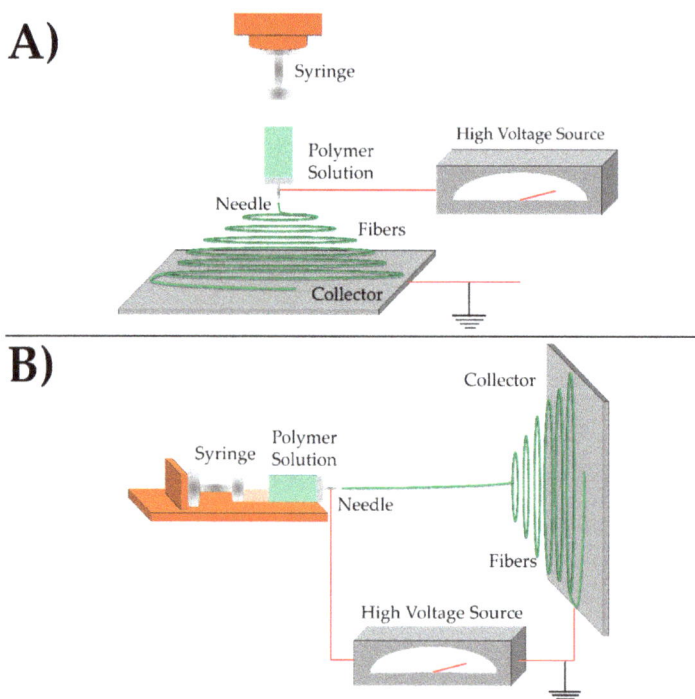

Figure 2. Schematic showing the set-up of a (**A**) vertical electrospinning system and (**B**) horizontal electrospinning system.

Before the electrospinning process, polymers are dissolved into a solvent and the solution is placed inside the syringe. To begin the process, the solution is forced out of the syringe at a constant flow rate. Simultaneously, a high voltage is applied to the solution, resulting in repulsive interactions between like charges within the solution. A Taylor cone [138] is formed at the end of the syringe when the electrical forces and surface tension forces in the solution are at equilibrium. At a critical value, the electrical forces overcome the surface tension forces and a jet of solution propels out of the Taylor cone and towards the grounded collector. At ideal conditions, as the solution jet travels, the solvent evaporates from the solution, leaving non-woven, polymer fibers due to high surface area to volume ratio, and finally gathered on the collector [131–137]. Fibers are produced with diameters in the range of 10 nm–10 μm, and various collector modifications can also allow the formation of aligned nanofiber arrays and non-woven yarns. In the literature, numerous protein nanofibers have been fabricated using the electrospinning technique. Generated fiber mats for silk, collagen, and gelatin-based fibers are shown in Figure 3 at various scales.

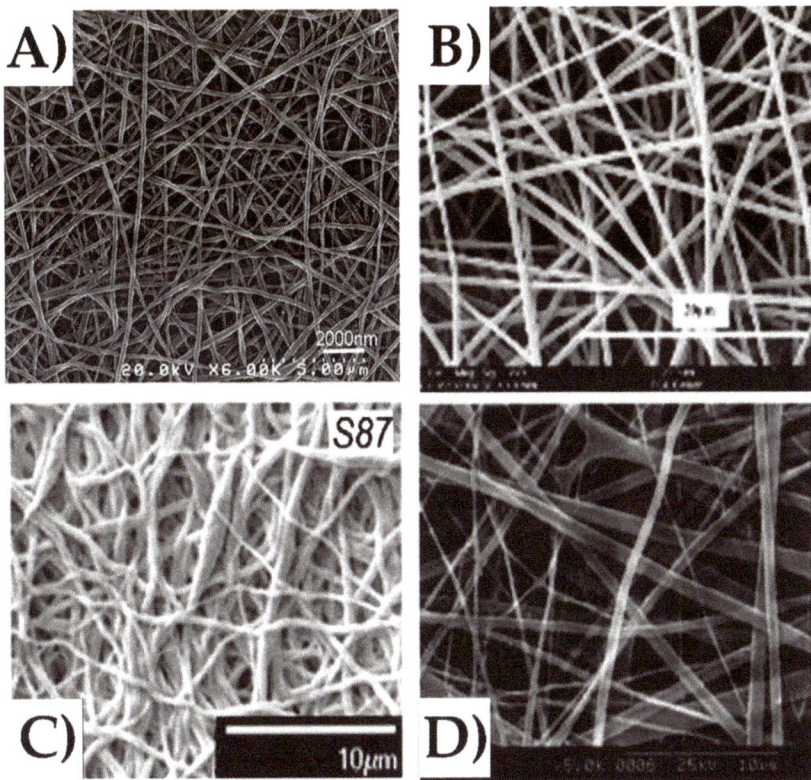

Figure 3. Scanning electron microscope images of (**A**) pure silk nanofibers, (**B**) PCL-gelatin nanofibers, (**C**) silk-PEO nanofibers, and (**D**) type I collagen nanofibers fabricated using the electrospinning technique (reproduced with permission from [19,78,136,139], copyright Elsevier, 2017 (**A**); Elsevier, 2007 (**B**); John Wiley and Sons, Inc., 2010 (**C**); Elsevier, 2006 (**D**)).

Electrospinning is a simple, reliable process that produces fibers with controllable properties. The process is able to produce a versatile range of fibers, including polymer-, synthetic-, and composite-based fibers [131,134]. The properties of the fibers can be influenced by controlling different parameters. These parameters can be categorized into three types: solution, process, and ambient [131,134]. Table 1 lists controllable parameters in their respective categories. Reproducibility and functionalization of protein-based nanofibers may be enhanced by treating the protein prior to fabrication. This approach was adopted by Pegg et al. [140] to produce alginic acid nanofibers. Prior to spinning, the alginic acid was converted to ammonium alginate by reacting the polymer with amine-containing cargo. This pre-treatment allowed for more uniform functionalization and enabled the fibers to carry diverse therapeutics, such as lidocaine, neomycin, and papain. Electrospun nanofibers like these are also useful in a variety of applications due to their high surface-area-to-volume ratio [131,134] and modifiable surface porosity [134].

3.2. Wet/Dry-Jet Spinning

Wet spinning is a fiber fabrication method that consists of a polymer solution, a spinneret, and a coagulation bath (Figure 4A). During the fabrication process, the polymer solution is extruded via a syringe pump through the spinneret directly into a coagulation bath. Polymer fibers form in the coagulation bath as the solvent is removed either through chemical reaction or diffusion [141]. After formation, the remaining fiber material is collected and dried. Drawing, or applying tension to the fibers can occur immediately after the spinneret [141], during drying [125], or further down the spinning line [142,143] to elongate the fiber, increase molecular alignment and, consequently, stiffness and strength. Wet spinning fabrication systems may implement multiple drawings or baths in order to improve molecular alignment and orientation.

A modified version of wet spinning, referred to as dry-jet wet spinning has been developed (Figure 4B) [124,141,142]. In dry-jet wet spinning, the polymer solution is extruded through an air gap before the coagulation bath, rather than directly into the bath. Studies have shown that dry-jet wet spinning can result in greater molecular alignment compared to conventional wet spinning [142].

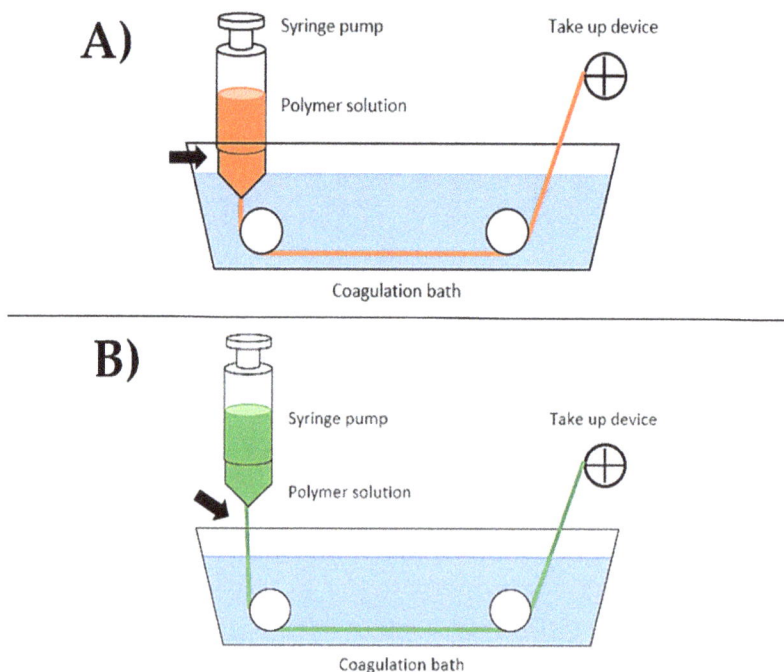

Figure 4. Common systems of (**A**) wet spinning, and (**B**) dry-jet wet spinning. In dry-jet wet spinning, the polymer solution is extruded through an air gap before the coagulation bath, resulting in higher molecular alignment compared to conventional wet spinning [124,141,142].

By controlling parameters, such as the diameter of the spinneret, polymer solution concentration, and flow rate, fiber properties, such as diameter, orientation, and morphology [122,125,141,143], can be modified. Unlike in electrospinning, the fibers are not exposed to a high voltage that may denature natural polymers, such as proteins. Additionally, drawing of the fibers after formation can lead to enhanced material properties due to higher molecular alignment. However, the wet/dry spinning methods typically produces only micron-sized fibers, while electrospinning is a common method of producing nanofibers.

3.3. Dry Spinning

Dry spinning is a fiber fabrication method that consists of a polymer solution, a syringe, and a collector. Figure 5A shows the scheme of a typical dry spinning system. Unlike other solution spinning methods, such as electrospinning and wet spinning, dry spinning involves a single extrusion step. A key part of the method is that the polymer/protein solution is created such that the solvent of the solution will evaporate in the ambient environment during spinning. During the process, the solution is pumped through a syringe and spinneret. Ideally, the solvent of the solution will evaporate out of the solution, leaving only the polymer fiber to be collected. The fibers are then collected via a take-up device similar to those used in the wet spinning processes [126]. Similar to wet spinning, additional drawing [144], heating, or drying [126] can occur to increase the mechanical properties or ensure fiber stability.

The mechanical properties of the fiber can be affected controlling take up speed, length to diameter ratio of the spinneret, environment or temperature, and spinning rate [126,145]. Additionally, subjecting the dry-spun fiber to a post-treatment agent or post-formation drawing can enhance mechanical properties [126].

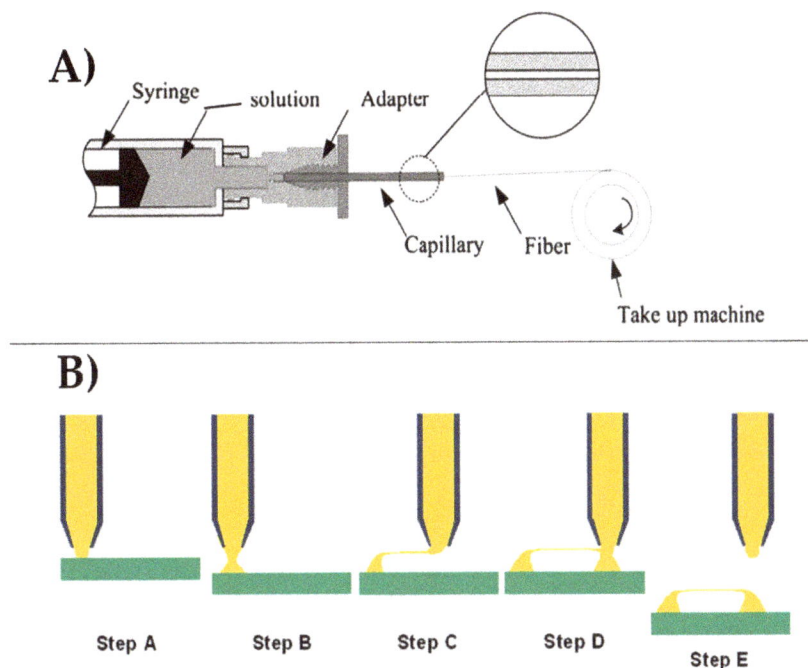

Figure 5. (**A**) Scheme of a typical dry spinning system. (**B**) Representation of drawing mechanism for polymer-based fiber fabrication (reproduced with permission from [126,146], copyright Elsevier, 2011 (**A**); AIP Publishing, 2006 (**B**)).

3.4. Centrifugal Spinning

Centrifugal spinning is a process commonly used in the industrial production of fiberglass. More recently the process has gained traction as a fabrication method of polymer fibers [129]. It exhibits some significant advantages over the more commonly practiced electrospinning method. Namely, its comparatively high production rate and its lack of dependence on a voltage resulting in greater safety [128,129]. A biopolymer solution or melt can be placed in a rotating head with a small opening referred to as the nozzle. When the head is rotated at a speed that exerts a centrifugal force on the solution or melt higher than its surface tension, the solution or melt will emerge from the nozzle as a liquid jet. The liquid jet is stretched by the combination of the centrifugal force and the air friction force and deposited into a collection area. Solidified fibers with diameters ranging from hundreds of nanometers to tens of micrometers are produced upon the evaporation of the solvent [129,147]. A process of nozzle-free centrifugal spinning has been tested by Weitz et al., resulting in fibers as small as 25 nm in diameter [148]. The viscosity, surface tension, molecular structure, molecular weight, polymer concentration, solvent structure, solvent evaporation rate, and additive of the polymer solution or melt all contribute to the morphology of the fibers, with viscosity and surface tension being the largest influences [128,129]. Rotational speed, head diameter, nozzle diameter, and distance between the nozzle and collector also largely influence the morphology of the fibers produced [129].

3.5. Solution Blowing

Solution blowing is a relatively new method of fiber fabrication. It involves an apparatus consisting of two concentric nozzles. A biopolymer solution is pumped through the inner nozzle as a high velocity gas flows through the outer nozzle. The flow of gas stretches the solution and ejects it from the apparatus. Fibers are formed in the air as the solvent evaporates before reaching a collector [149]. The high velocity gas is supplied from a source of compressed gas such as nitrogen, argon, or air equipped with a pressure regulator and connected to the apparatus. Biopolymer type, concentration, injection rate, gas flow pressure, and working distance all influence the properties of the fibers being produced [130].

3.6. Self-Assembly

Molecular self-assembly is a process ubiquitous in natural biological systems. Structures formed by self-assembly are governed by non-covalent forces, such as hydrogen bonding, electrostatic interactions, van der Waals interactions, hydrophobic interactions, stacking interactions, and water-mediated hydrogen bonding [150,151]. These non-covalent bonds between small molecules result in supramolecular architectures, such as nanofibers. The shapes and properties of resulting fibers are determined by the molecules and non-covalent bonds structuring them [152]. The driving intermolecular interactions can be influenced by environmental factors, such as salt concentration, pH, temperature, and surface characteristics [153].

3.7. Phase Separation

Phase separation is a rather simple process. However, it is limited to the scale of a laboratory setting [152]. The process begins with the dissolution of a protein polymer in a solvent. The temperature of the solution is then reduced to the gelation temperature, which is the point at which the solution forms a gel. Solvent exchange is carried out by immersing the gel in distilled water. When removed, the gel is blotted with filter paper and freeze-dried, resulting in the formation of a nanofibrous matrix [152]. By adjusting factors, such as gelation temperature and biopolymer concentration, the morphology of the fibers is able to be controlled [154].

3.8. Drawing

The commonly used fabrication method, electrospinning, produces a layer of fibers on a flat collector [155], while the fabrication method known as drawing only produces one fiber at a time. The production of single fibers limits the use of this method to the laboratory scale [152]. In the drawing process, a sharp-tipped probe is placed in contact with a droplet of biopolymer solution and withdrawn at a predetermined speed. The solvent evaporates due to the high surface area in relation to the volume. The end of the resulting fiber attached to the probe can then be connected to another droplet to form a suspended fiber [146]. The drawing process relies on the viscoelasticity of the solution so that it can maintain cohesion under the stresses of being pulled [156]. Multiple fibers can be drawn from each droplet [152]. When too much time is allowed to pass between the deposition of the solution droplet and the drawing of the fiber, the droplet will become too viscous due to the evaporation of the solvent. Furthermore, the continual shrinkage of the droplet affects the diameter of the fibers produced and limits the continuous drawing of fibers. A modification to this method, implementing the use of micropipettes, can improve the continuous formation of short fibers and provide greater control of the parameters that affect fiber properties. In this method the solution is continuously pumped through the micropipette. The droplet is formed at the tip of the micropipette and brought in contact with a substrate that it will adhere to. It is then laterally drawn before coming in contact with the substrate once again, forming a suspended fiber [146].

4. Applications of Protein-Based Nanofibers in Tissue Regeneration and Nanomedicine

Many protein-based fiber materials have applications in the biomedical field. These materials have been put to use, in part, because of their favorable biocompatible and biodegradable characteristics. For instance, natural protein fibers, such as keratin, silks, or collagens, are all of interest to researchers due to their mechanical properties and ability to maintain a low host immune response [157]. Protein-based fiber materials are often used for scaffolds, sutures, wound healing, ligament replacement, and drug delivery technology. Examples of these applications are summarized in Table 2 at the end of this section.

4.1. Tissue Engineering and Regenerative Medicine

Tissue engineering has become a major focus in the biomedical engineering community due to the lack of tissue transplants and host rejection of foreign tissue [158]. To be successful, two components must be optimized in tissue engineering—the cells and the scaffold. The scaffold is necessary as it provides the specific architecture and mechanical structure of the desired tissue by closely recapitulating the natural ECM [158,159].

Protein-based nanofiber membranes can provide an excellent scaffold in tissue engineering due to the biocompatibility, biodegradability, and tunability of the fibers. The presence of innate cell adhesion sites and functionality of these constructs also suggests the superiority of protein-based materials over those made from synthetic polymers when creating scaffolds for tissue engineering. These scaffolds create a platform for seeding cells in a defined structure, such that it mimics the host morphology, to catalyze the growth of new specialized tissues. Fibrous membranes also allow for the development of porous scaffolds, essential for cell migration, gas exchange, diffusion of nutrients, cell communication, and the elimination of waste, enhancing of the growth of the native ECM and the proliferation of surrounding cells [159]. Figure 6 maps out the different areas of the human body that have a need for tissue engineering or regenerative medicine applications.

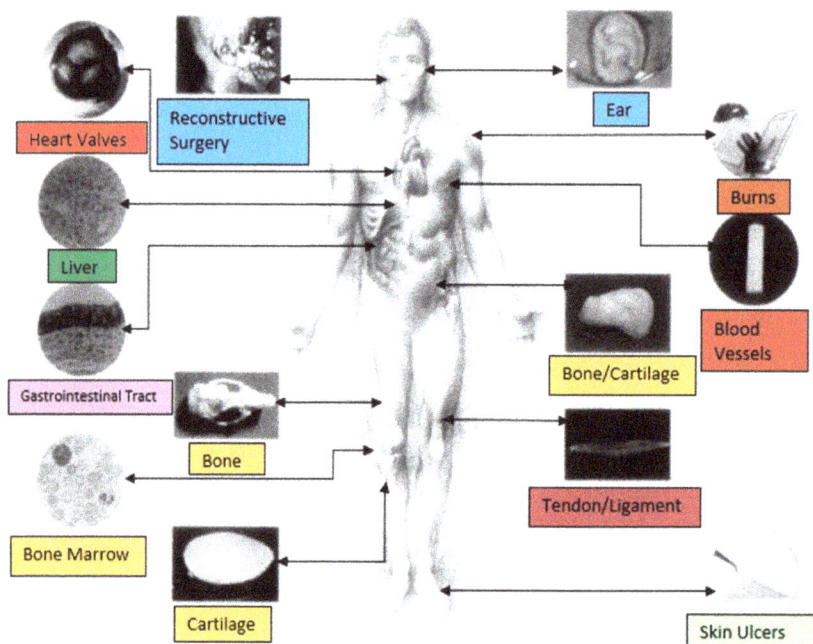

Figure 6. Generalized model of potential tissue engineering and medicine applications for various organ systems. (Reproduced with permission from [158], Copyright John Wiley and Sons, Inc., 1998).

Due to its abundance in the ECM, collagen fibers are most commonly used for the creation of tissue engineering scaffolds. Studies have successfully used these constructs for 3D cell culturing, vascular regeneration, skin grafts, bone tissue engineering, cartilage repair, nerve regeneration, spinal cord healing, and corneal defect correction [160–166]. For example, using the electrospinning method, Ribeiro et al. [167] developed collagen nanofiber meshes with an average fiber diameter of 30 nm. During spinning, nanohydroxyapatite crystals were also deposited onto the fibers by simultaneous electrospraying. Nanofibers were crosslinked with N-ethyl-N'-(3-dimethylaminopropyl) carbodiimide/N-hydroxy succinimide, allowing for the creation of a scaffold that closely recapitulated the native ECM of bone tissue and allowed for osteoblast adhesion and proliferation.

Collagen-based nanofibers have also been co-fabricated with synthetic polymers to enhance the mechanical properties of the scaffolds. For example, as seen in Figure 7A,B, Tillman et al. created an electrospun PCL-collagen scaffold for a rabbit aortoiliac bypass [168]. The scaffold supported cell growth and was able to withstand normal physiological conditions. Additionally, it supported adhesion and growth of vascular cells, which was important for nutrient delivery and functionality of the implanted scaffold. Lastly, it maintained its structural integrity for over one month during the experiment. Once it was removed, the scaffold displayed biomechanical strength, comparable to its intended native artery. Ekaputura et al. also used similar collagen-PCL composite nanofibers encased in a hyaluronic acid hydrogel to promote vascularized bone regeneration through the release of vascular endothelial growth factor and platelet-derived growth factor [169].

Often, synthetic polymers are incorporated into protein-based fibrous materials to improve mechanical stability. However, proteins, such as elastin, have been incorporated into synthetic polymer-based scaffolds to modulate their mechanical properties. Foraida et al. [18] covalently conjugated elastin onto the surface of electrospun poly-lactic-co-glycolic acid (PLGA) nanofiber scaffolds through 1-ethyl-3-(3-dimethylaminopropyl) carbodiimide/N-hydroxysulfosuccinimide

(EDC/NHS) chemistry. While this was found to improve the wettability of the scaffolds, it had little effect on their elasticity. Therefore, elastin and PLGA were also blended prior to electrospinning to create elastin-PLGA composite meshes containing fibers with an average diameter of approximately 300 nm. The incorporation of elastin was found to greatly increase the compliance of the fiber meshes. Compared to PLGA fibers with a Young's Modulus of 4.29 MPa, the modulus of elastin-PLGA fibers was 0.59 MPa. As a result, elastin-PLGA nanofibers were able to support apical polarization and self-organization of epithelial cells, allowing for controllable cell proliferation and a higher degree of cell-cell contact compared to PLGA fibers. These characteristics better recapitulate the native arrangement of epithelial cells. As a result, elastin was identified as an integral part of tissue engineering scaffolds that hope to promote regeneration of epithelial constructs, like salivary glands.

In addition to collagen and elastin, fibrous proteins, such as silk, have been used to create tissue engineering scaffolds. Figure 7C–E depicts an in vivo rat study reported on by Melke et al. to assess the capacity of scaffolds made from mulberry *B. mori* silk and non-mulberry *A. mylitta* to induce bone regeneration in a cranial defect model. Since mylitta silk contains the natural RGD (Arg-Gly-Asp) motif, it allowed for enhanced cell adhesion and proliferation, leading to the regeneration of a higher bone volume [170]. Kim et al. has also used silk nanofiber meshes to induce bone regeneration. These meshes were also seeded with stem cells and evaluated after 31 days. They found their stem cell-seeded scaffolds and meshes were able to guide differentiation and promote new bone formation [171].

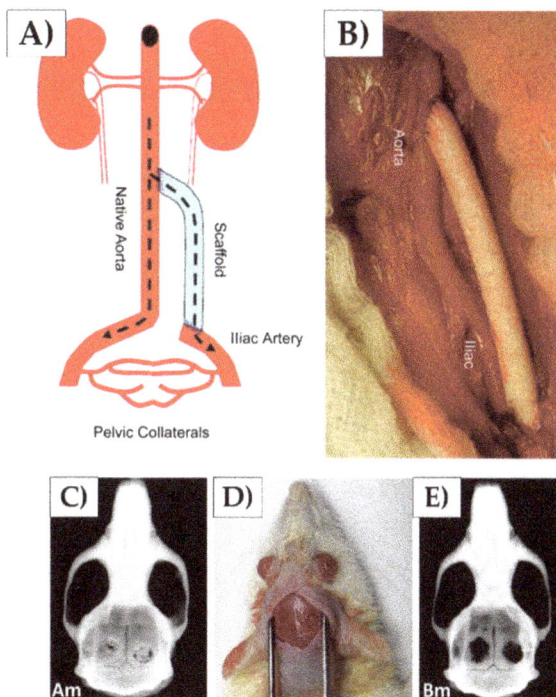

Figure 7. (**A**,**B**) Aortoiliac bypass using a PCL-collagen scaffold; (**C**–**E**) Effects of mulberry *B. mori* silk and non-mulberry *A. mylitta* silk on bone regeneration (reproduced with permission from [168,170], Copyright Elsevier, 2009 (**A**,**B**); Elsevier 2016 (**C**–**E**)).

Due to its mechanical strength, silk is an ideal protein for tissue regeneration. Recently, Du et al. [87] also used silk fibroin to produce nanofibrous scaffolds for heart valve tissue engineering, but combined the natural polymer with poly(ester-urethane) (PEUU) to improve the fracture resistance of the scaffold. These scaffolds were created by combining the two polymers in hexafluoroisopropanol and following the electrospinning technique previously highlighted. Meshes consisting of randomly-orientated fibers with varying diameters were created. Average diameter and the hydrophilicity of the scaffold were dependent upon the ratio of silk fibroin to PEUU, both of which decreased with the addition of PEUU. However, these changes did not significantly affect cell adhesion and proliferation, illustrating that the mechanical properties of silk materials can be easily optimized through the addition of synthetic polymers without sacrificing biocompatibility.

Nanofibrous scaffolds made from gelatin have also been developed and may have a promising future in cartilage tissue regeneration. Agheb et al. [10] developed electrospun gelatin fiber meshes that were crosslinked before or after synthesis with glutaraldehyde or EDC/NHS chemistry. To increase the functionality of these scaffolds, the authors also embedded tyrosine and triazole rings since an increase in aromatic ring content has also been shown to enhance the tensile strength and bioactivity of materials. Fibers crosslinked after electrospinning exhibited larger diameters, reduced porosity, greater rigidity, and smaller pore sizes. As a result, those crosslinked before electrospinning sustained better chondrocyte proliferation and viability in later tests. The incorporation of additional aromatic rings was also shown to increase chondrocyte viability and allowed them to express their natural phenotype and morphology, leading the authors to believe that the modified gelatin scaffolds could be used to promote cartilage regeneration in vivo.

These above studies illustrate that protein-based nanofibers can be used to promote regeneration of a variety of tissue in nanoscale. In all cases, nanofibers were easily fabricated and modified by chemical functionalization, crosslinking, or polymer blending showcasing the versatility of such materials.

4.2. Drug Delivery

Nanofiber materials are often used in drug delivery since their high surface-area-to-volume ratio and porosity allow for high efficiency drug storage and release. Protein-based nanofibers have particular appeal over their synthetic counterparts since the materials are biodegradable and biocompatible. They have also been shown to be highly modifiable, allowing researchers to tune the release of pharmaceuticals.

Due to the biodegradability, flexibility, biocompatibility, anti-microbial properties, and the anti-oxidant behavior of corn zein protein, it is commonly used to create nanofibers that are used in drug delivery [19,172]. Figure 8 shows the results of a drug delivery study done with co-axial electrospun corn zein nanofibers [173]. In coaxial electrospinning, two liquids are spun simultaneously from one concentric spinneret to create fibers with a core-sheath structure. However, to generate higher quality fibers with increased smoothness and uniformity, an unspinnable solvent containing no polymer was used to create the sheath, while corn zein was used to create the core. This unspinnable solvent, in this case dimethylformamide (DMF), is unable to form nanofibers, but replaces the standard polymer-air interface usually seen during electrospinning. In this way, an air-solvent-polymer interface is established, thus mitigating environmental effects on fiber formation. Ferulic acid (FA) was incorporated into the corn zein-ethanol solution prior to spinning and served as the model drug. Corn zein nanofibers created through standard electrospinning (without the unspinnable solvent sheath) were designated F1, while those made with coaxial spinning were designated F2 (Figure 8). As seen in Figure 8A, FA is released from both types of electrospun corn zein fibers through a standard Fickian diffusion mechanism. However, F2 fibers exhibited a longer, more controlled release than F1. This was attributed to the fact that F1 fibers had a flatter, more ribbon-like structure than F2 due to the coaxial spinning mechanism. F1 fibers also had a more wrinkled appearance than F2. Both of these characteristics contribute to a higher surface area on the F1 fibers, leading to faster, burst release of FA. This study illustrates the use of corn zein nanofibers as drug delivery vehicles and also the ability to optimize fiber fabrication methods to achieve the desired release profiles of drugs.

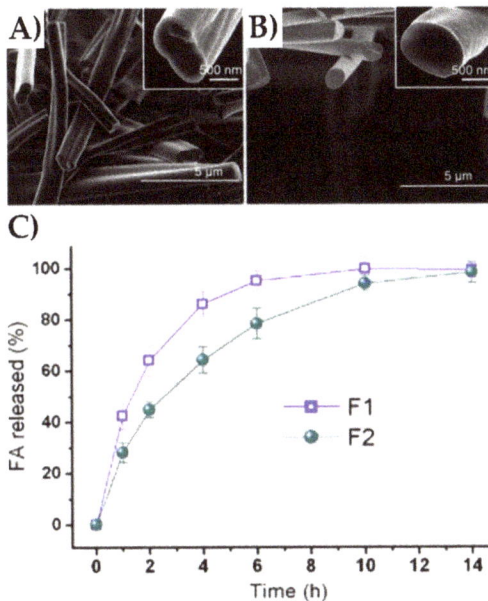

Figure 8. Corn zein nanofibers containing ferulic acid (FA) can be fabricated through single-fluid electrospinning (**A**) and modified coaxial electrospinning (**B**). Release of FA from both fiber types (F1 and F2) was then monitored over time (**C**) (reproduced with permission from [173], Copyright Elsevier, 2013).

Soy protein has also been used to create fiber meshes for applications in drug delivery. Xu et al. [174] developed soy protein microfibers by dissolving the protein in an aqueous urea solution and extruding it into a sodium sulfate solution. This produced fibers with an average diameter of 45 µm. Model drugs, including diclofenac, 5 fluorouracil (5-Fu), and metformin, were incorporated into fibers prior to spinning or loaded into fibers through sorption by exposing the dry fibers to solutions containing drugs at various conditions. Overall, the researchers found that the soy protein fibers had a high affinity to the model drugs, allowing for efficient sorption loading. This loading could also be modified by modulating the temperature during fiber exposure. Burst release could also be limited by lowering the concentration of the loaded drug.

Like those used in tissue engineering, natural and synthetic polymer composite fibers can also be used for drug release. For example, Lee et al. [175] developed multi-layered PLGA/collagen nanofibers membranes through electrospinning, that were used to deliver lidocaine and epinephrine over four weeks. These membranes were then used to deliver the drugs to rabbits with palatal oral wounds and test groups showed faster hemostasis, as well as recovery of food and water intake, compared to control groups who received an empty PCL membrane. In place of a model drug, proteins have also been added to synthetic polymer-based nanofibers and released over time. Zeng et al. [176] prepared these protein-loaded fibers by electrospinning a solution of poly(vinyl alcohol) and bovine serum albumin protein (BSA). To control the release of BSA, the fibers were also coated with poly(p-xylylene) (PPX). This coating was found to slow the release of BSA over 20 days, and successful release of the model enzyme luciferase was also observed. This demonstrates that proteins can be easily incorporated into fibers for structural purposes, as well as therapeutic ones.

4.3. Wound Healing

Due to their porosity, gas permeability, and high surface-area-to-volume ratio, fibrous materials may offer advanced treatment options for burn victims and patients with skin ulcers compared to conventional treatment options like hyperbaric oxygen therapy. A successful wound dressing is one that is able to facilitate epithelial cell migration and regeneration, and this is often best achieved by creating a warm, moist environment. During healing, it is also important to prevent the influx of bacteria that can lead to an infection that delays wound healing. To meet these requirements, protein-based fiber meshes are commonly used for the creation of conventional bandages, antimicrobial-infused dressings, and advanced tissue engineered skin grafts [177,178].

In an effort to promote wound healing and minimize infection and inflammation Chouhan et al. [25] recapitulated the physical and biological ECM of dermal tissue by creating electrospun silk fibroin meshes with varying concentrations of poly(vinyl alcohol) (PVA) for added mechanical stability. Silk fibroin from the *Bombyx mori* (PVAABM) silkworm was used, as well as non-mulberry silk from *Antheraea assama* (PVAAA) and *Philosamia ricini* (PVAPR). These meshes were then functionalized with epidermal growth factor (EGF) and ciprofloxacin HCl antibiotic. All fibers allowed burst release of EGF, but those containing silk fibroin achieved a greater release over 24 h, compared to those made solely from PVA. Since non-mulberry silk fibroin contains a naturally occurring RGD motif, these fiber meshes were also able to support greater cell adhesion and proliferation, as seen in Figure 9A. Silk fibroin also allowed for greater water retention, allowing for the creation of the moist environment, crucial in wound regeneration. Due to these characteristics, nanofiber meshes containing silk fibroin resulted in faster wound closure than meshes containing PVA alone, as seen in Figure 9B,C. Meshes containing non-mulberry silk fibroin out-performed those containing *B. mori* silk fibroin with 100% closure being achieved after 14 days, compared to only 80% closure in the latter samples.

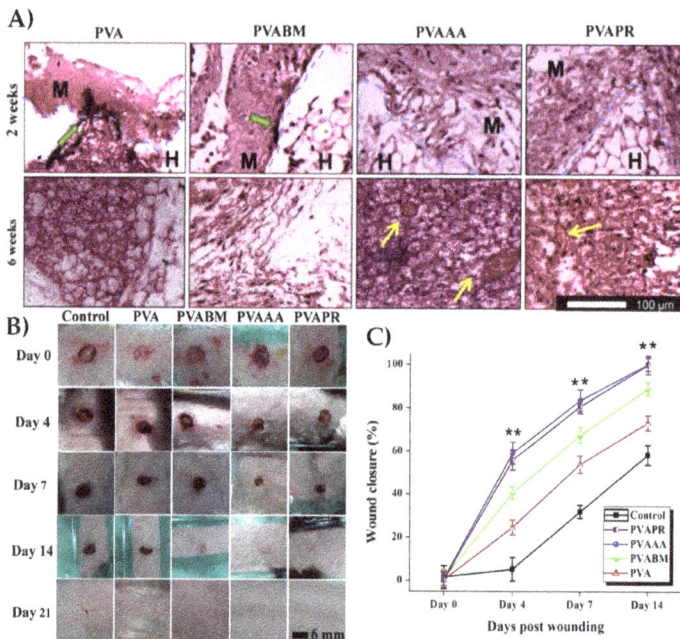

Figure 9. (**A**) Hematoxylin and eosin staining for histological analysis of subcutaneously implanted mats made of varying PVA-silk nanofiber composite fibers. Mats containing non-mulberry silk showed greater cell infiltration and proliferation compared to those made of mulberry silk or those containing only PVA. Green arrows indicate infiltrating cells at the interface of the implant, and yellow arrows indicate the development of new blood vessels. (**B**) Images of wounds on rabbits treated with PVAA-silk mats. The percentage of wound closure is quantified in (**C**) (reproduced with permission from [25], Copyright Elsevier, 2017).

In addition to growth factors and antibiotics, nanoparticles can also be embedded into protein-based fiber meshes. This is advantageous, since some metal nanoparticles, such as those made from silver (AgNPs), can have antimicrobial properties. Wang et al. [179] incorporated AgNPs into keratin-based nanofibers that were blended with polyurethane and fabricated through the electrospinning process. When used to heal circular wounds in rats, these nanofiber meshes resulted in 30% wound closure after nine days, compared to only 60% wound closure in a control group that received conventional sponge dressing. A reduction of TNF-α secretion and inflammatory cell infiltration was also seen. The incorporation of keratin into poly(hydroxybutylate-cohydroxyvalerate) (PHBV) nanofiber mats was also shown to accelerate the proliferation of human fibroblast cells, as demonstrated by Yuan et al. [180].

Table 2. Overview of applications of protein-based nanofibers in medicine.

Protein	Fabrication Technique	Material	Application
Keratin	Electrospinning	• AGNP-embedded keratin-Polyurethane nanofibers • PHBV-keratin nanofiber mats	• Dermal wound healing [179] • Fibroblast cell proliferation [180]
Collagen	Electrospinning	• Crosslinked collagen nanofiber meshes • PCL-collagen nanofiber meshes • PCL-collagen nanofiber meshes embedded in hyaluronic acid hydrogel • PLGA-collagen nanofibers	• Bone tissue regeneration [167] • Aortoiliac bypass [168] • Vascularized bone regeneration [169] • Delivery of lidocaine and epinephrine [175]
Gelatin	Electrospinning	• Crosslinked gelatin nanofibers with embedded tyrosine and triazole rings	• Cartilage regeneration [10]
Silk Fibroin	Electrospinning	• Silk fibroin nanofiber meshes • PEUU-silk fibroin nanofiber mashes • EGF-functionalized PVA-silk fibroin nanofibers	• Bone tissue regeneration [171] • Heart valve regeneration [87] • Dermal wound healing [25]
Zein	Co-axial Electrospinning	• Corn zein nanofibers	• Delivery of ferulic acid [173]
Soy	Wet-spinning	• Soy protein microfibers	• Delivery of diclofenac, 5 fluorouracil and metformin [174]
Elastin	Electrospinning	• Crosslinked PLGA-elastin nanofiber meshes	• Regeneration of epithelial constructs [18]

5. Conclusions

The use of fibrous materials in biomedical research has been growing in popularity due to their high surface-area-to-volume ratio, tunability, porosity, and mechanical strength. Fibers made from natural polymers, such as proteins, hold additional promise due to their biodegradability and biocompatibility. Natural enzymes within the body are able to degrade these proteins to produce amino acids that pose no risk of toxicity and can be reabsorbed. Proteins, such as silk, collagen, and keratin, also contain innate cell adhesion motifs. Therefore, when these constructs are used in tissue engineering or wound healing applications, they can support increased cell migration and proliferation. The mechanical and physical properties of many protein-based fiber materials can also be modified by crosslinking or blending an additional polymer, creating a tunable platform. These characteristics are also highly dependable on the type of the fabrication method used to create fibers. While electrospinning is currently the most popular technique, other methods, such as wet/dry jet spinning, dry spinning, centrifugal spinning, solution blowing, self-assembly, phase separation, and drawing, have been successful. As biomedical research and technology progresses, protein-based fibers may lead the way in the development of new biomaterials that promote tissue regeneration, wound healing, and controllable drug delivery.

Author Contributions: K.G.D., R.M., J.B., G.L., T.M., V.B., and X.H. wrote the paper together.

Funding: This study was supported by Rowan University Seed Research Grants, and the US NSF Materials Eng. and Processing Program (CMMI-1561966).

Conflicts of Interest: The authors declare no conflict of interest.

References

1. Sang, Q.; Williams, G.R.; Wu, H.; Liu, K.; Li, H.; Zhu, L.-M. Electrospun gelatin/sodium bicarbonate and poly(lactide-*co*-ε-caprolactone)/sodium bicarbonate nanofibers as drug delivery systems. *Mater. Sci. Eng. C* **2017**, *81*, 359–365. [CrossRef] [PubMed]

2. Löbmann, K.; Svagan, A.J. Cellulose nanofibers as excipient for the delivery of poorly soluble drugs. *Int. J. Pharm.* **2017**, *533*, 285–297. [CrossRef] [PubMed]

3. Lv, Y.; Pan, Q.; Bligh, S.W.A.; Li, H.; Wu, H.; Sang, Q.; Zhu, L.-M. Core-Sheath Nanofibers as Drug Delivery System for Thermoresponsive Controlled Release. *J. Pharm. Sci.* **2017**, *106*, 1258–1265. [CrossRef] [PubMed]

4. Vashisth, P.; Raghuwanshi, N.; Srivastava, A.K.; Singh, H.; Nagar, H.; Pruthi, V. Ofloxacin loaded gellan/PVA nanofibers—Synthesis, characterization and evaluation of their gastroretentive/mucoadhesive drug delivery potential. *Mater. Sci. Eng. C* **2017**, *71*, 611–619. [CrossRef] [PubMed]

5. Mutlu, G.; Calamak, S.; Ulubayram, K.; Guven, E. Curcumin-loaded electrospun PHBV nanofibers as potential wound-dressing material. *J. Drug Deliv. Sci. Technol.* **2018**, *43*, 185–193. [CrossRef]

6. Waghmare, V.S.; Wadke, P.R.; Dyawanapelly, S.; Deshpande, A.; Jain, R.; Dandekar, P. Starch based nanofibrous scaffolds for wound healing applications. *Bioact. Mater.* **2017**. [CrossRef] [PubMed]

7. Tort, S.; Acartürk, F.; Beşikci, A. Evaluation of three-layered doxycycline-collagen loaded nanofiber wound dressing. *Int. J. Pharm.* **2017**, *529*, 642–653. [CrossRef] [PubMed]

8. Zhang, H.; Xia, J.; Pang, X.; Zhao, M.; Wang, B.; Yang, L.; Wan, H.; Wu, J.; Fu, S. Magnetic nanoparticle-loaded electrospun polymeric nanofibers for tissue engineering. *Mater. Sci. Eng. C* **2017**, *73*, 537–543. [CrossRef] [PubMed]

9. Rijal, N.P.; Adhikari, U.; Khanal, S.; Pai, D.; Sankar, J.; Bhattarai, N. Magnesium oxide-poly(ε-caprolactone)-chitosan-based composite nanofiber for tissue engineering applications. *Mater. Sci. Eng. B* **2018**, *228*, 18–27. [CrossRef]

10. Agheb, M.; Dinari, M.; Rafienia, M.; Salehi, H. Novel electrospun nanofibers of modified gelatin-tyrosine in cartilage tissue engineering. *Mater. Sci. Eng. C* **2017**, *71*, 240–251. [CrossRef] [PubMed]

11. Unal, B.; Yalcinkaya, E.E.; Demirkol, D.O.; Timur, S. An electrospun nanofiber matrix based on organo-clay for biosensors: PVA/PAMAM-Montmorillonite. *Appl. Surf. Sci.* **2018**, *444*, 542–551. [CrossRef]

12. Guler Gokce, Z.; Akalın, P.; Kok, F.N.; Sarac, A.S. Impedimetric DNA biosensor based on polyurethane/poly(m-anthranilic acid) nanofibers. *Sens. Actuators B Chem.* **2018**, *254*, 719–726. [CrossRef]

13. Migliorini, F.L.; Sanfelice, R.C.; Mercante, L.A.; Andre, R.S.; Mattoso, L.H.C.; Correa, D.S. Urea impedimetric biosensing using electrospun nanofibers modified with zinc oxide nanoparticles. *Appl. Surf. Sci.* **2018**, *443*, 18–23. [CrossRef]

14. Liang, D.; Hsiao, B.S.; Chu, B. Functional electrospun nanofibrous scaffolds for biomedical applications. *Adv. Drug Deliv. Rev.* **2007**, *59*, 1392–1412. [CrossRef] [PubMed]

15. Leung, V.; Ko, F. Biomedical applications of nanofibers. *Polym. Adv. Technol.* **2011**, *22*, 350–365. [CrossRef]

16. Bhattarai, N.; Edmondson, D.; Veiseh, O.; Matsen, F.A.; Zhang, M. Electrospun chitosan-based nanofibers and their cellular compatibility. *Biomaterials* **2005**, *26*, 6176–6184. [CrossRef] [PubMed]

17. Elakkiya, T.; Malarvizhi, G.; Rajiv, S.; Natarajan, T.S. Curcumin loaded electrospun Bombyx mori silk nanofibers for drug delivery. *Polym. Int.* **2014**, *63*, 100–105. [CrossRef]

18. Foraida, Z.I.; Kamaldinov, T.; Nelson, D.A.; Larsen, M.; Castracane, J. Elastin-PLGA hybrid electrospun nanofiber scaffolds for salivary epithelial cell self-organization and polarization. *Acta Biomater.* **2017**, *62*, 116–127. [CrossRef] [PubMed]

19. Babitha, S.; Rachita, L.; Karthikeyan, K.; Shoba, E.; Janani, I.; Poornima, B.; Purna Sai, K. Electrospun protein nanofibers in healthcare: A review. *Int. J. Pharm.* **2017**, *523*, 52–90. [CrossRef] [PubMed]

20. Ulery, B.D.; Nair, L.S.; Laurencin, C.T. Biomedical Applications of Biodegradable Polymers. *J. Polym. Sci. Part B Polym. Phys.* **2011**, *49*, 832–864. [CrossRef] [PubMed]

21. Vasconcelos, A.; Freddi, G.; Cavaco-Paulo, A. Biodegradable Materials Based on Silk Fibroin and Keratin. *Biomacromolecules* **2008**, *9*, 1299–1305. [CrossRef] [PubMed]

22. Zhang, H.; Zhou, L.; Zhang, W. Control of Scaffold Degradation in Tissue Engineering: A Review. *Tissue Eng. Part B Rev.* **2014**, *20*, 492–502. [CrossRef] [PubMed]

23. Karthikeyan, K.; Guhathakarta, S.; Rajaram, R.; Korrapati, P.S. Electrospun zein/eudragit nanofibers based dual drug delivery system for the simultaneous delivery of aceclofenac and pantoprazole. *Int. J. Pharm.* **2012**, *438*, 117–122. [CrossRef] [PubMed]

24. Vega-Lugo, A.-C.; Lim, L.-T. Controlled release of allyl isothiocyanate using soy protein and poly(lactic acid) electrospun fibers. *Food Res. Int.* **2009**, *42*, 933–940. [CrossRef]

25. Chouhan, D.; Chakraborty, B.; Nandi, S.K.; Mandal, B.B. Role of non-mulberry silk fibroin in deposition and regulation of extracellular matrix towards accelerated wound healing. *Acta Biomater.* **2017**, *48*, 157–174. [CrossRef] [PubMed]

26. Zhang, X.; Tang, K.; Zheng, X. Electrospinning and Crosslinking of COL/PVA Nanofiber-microsphere Containing Salicylic Acid for Drug Delivery. *J. Bionic Eng.* **2016**, *13*, 143–149. [CrossRef]

27. Hofmann, S.; Wong Po Foo, C.T.; Rossetti, F.; Textor, M.; Vunjak-Novakovic, G.; Kaplan, D.L.; Merkle, H.P.; Meinel, L. Silk fibroin as an organic polymer for controlled drug delivery. *J. Control. Release* **2006**, *111*, 219–227. [CrossRef] [PubMed]

28. Hu, X.; Cebe, P.; Weiss, A.S.; Omenetto, F.; Kaplan, D.L. Protein-based composite materials. *Mater. Today* **2012**, *15*, 208–215. [CrossRef]

29. DeFrates, K.; Markiewicz, T.; Callaway, K.; Xue, Y.; Stanton, J.; Salas-de la Cruz, D.; Hu, X. Structure–property relationships of Thai silk–microcrystalline cellulose biocomposite materials fabricated from ionic liquid. *Int. J. Biol. Macromol.* **2017**, *104*, 919–928. [CrossRef] [PubMed]

30. Song, F.; Tang, D.-L.; Wang, X.-L.; Wang, Y.-Z. Biodegradable Soy Protein Isolate-Based Materials: A Review. *Biomacromolecules* **2011**, *12*, 3369–3380. [CrossRef] [PubMed]

31. Kundu, B.; Rajkhowa, R.; Kundu, S.C.; Wang, X. Silk fibroin biomaterials for tissue regenerations. *Adv. Drug Deliv. Rev.* **2013**, *65*, 457–470. [CrossRef] [PubMed]

32. Davidenko, N.; Schuster, C.F.; Bax, D.V.; Farndale, R.W.; Hamaia, S.; Best, S.M.; Cameron, R.E. Evaluation of cell binding to collagen and gelatin: A study of the effect of 2D and 3D architecture and surface chemistry. *J. Mater. Sci. Mater. Med.* **2016**, *27*, 148. [CrossRef] [PubMed]

33. Pina, S.; Oliveira, J.M.; Reis, R.L. Natural-Based Nanocomposites for Bone Tissue Engineering and Regenerative Medicine: A Review. *Adv. Mater.* **2015**, *27*, 1143–1169. [CrossRef] [PubMed]

34. Brennan, D.; Jao, D.; Siracusa, M.; Wilkinson, A.; Hu, X.; Beachley, V. Concurrent collection and post-drawing of individual electrospun polymer nanofibers to enhance macromolecular alignment and mechanical properties. *Polymer* **2016**, *103*, 243–250. [CrossRef]

35. Chomachayi, M.D.; Solouk, A.; Mirzadeh, H. Electrospun silk-based nanofibrous scaffolds: Fiber diameter and oxygen transfer. *Prog. Biomater.* **2016**, *5*, 71–80. [CrossRef] [PubMed]

36. Daamen, W.F.; Veerkamp, J.H.; van Hest, J.C.M.; van Kuppevelt, T.H. Elastin as a biomaterial for tissue engineering. *Biomaterials* **2007**, *28*, 4378–4398. [CrossRef] [PubMed]

37. Wagenseil Jessica, E.; Mecham Robert, P. New insights into elastic fiber assembly. *Birth Defects Res. Part C Embryo Today Rev.* **2008**, *81*, 229–240. [CrossRef] [PubMed]

38. Daamen, W.F.; Nillesen, S.T.; Wismans, R.G.; Reinhardt, D.P.; Hafmans, T.; Veerkamp, J.H.; van Kuppevelt, T.H. A biomaterial composed of collagen and solubilized elastin enhances angiogenesis and elastic fiber formation without calcification. *Tissue Eng. Part A* **2008**, *14*, 349–360. [CrossRef] [PubMed]

39. Han, J.; Lazarovici, P.; Pomerantz, C.; Chen, X.; Wei, Y.; Lelkes, P.I. Co-Electrospun Blends of PLGA, Gelatin, and Elastin as Potential Nonthrombogenic Scaffolds for Vascular Tissue Engineering. *Biomacromolecules* **2011**, *12*, 399–408. [CrossRef] [PubMed]

40. Annabi, N.; Mithieux, S.M.; Camci-Unal, G.; Dokmeci, M.R.; Weiss, A.S.; Khademhosseini, A. Elastomeric Recombinant Protein-based Biomaterials. *Biochem. Eng. J.* **2013**, *77*, 110–118. [CrossRef] [PubMed]

41. Kim, W.; Chaikof, E.L. Recombinant Elastin-Mimetic Biomaterials: Emerging Applications in Medicine. *Adv. Drug Deliv. Rev.* **2010**, *62*, 1468–1478. [CrossRef] [PubMed]

42. Koria, P.; Yagi, H.; Kitagawa, Y.; Megeed, Z.; Nahmias, Y.; Sheridan, R.; Yarmush, M.L. Self-assembling elastin-like peptides growth factor chimeric nanoparticles for the treatment of chronic wounds. *Proc. Natl. Acad. Sci. USA* **2011**, *108*, 1034–1039. [CrossRef] [PubMed]

43. MacEwan Sarah, R.; Chilkoti, A. Elastin-like polypeptides: Biomedical applications of tunable biopolymers. *Pept. Sci.* **2010**, *94*, 60–77. [CrossRef] [PubMed]

44. Nuhn, H.; Klok, H.-A. Secondary Structure Formation and LCST Behavior of Short Elastin-Like Peptides. *Biomacromolecules* **2008**, *9*, 2755–2763. [CrossRef] [PubMed]

45. Xia, X.-X.; Xu, Q.; Hu, X.; Qin, G.; Kaplan, D.L. Tunable Self-Assembly of Genetically Engineered Silk–Elastin-like Protein Polymers. *Biomacromolecules* **2011**, *12*, 3844–3850. [CrossRef] [PubMed]

46. McClure, M.J.; Sell, S.A.; Simpson, D.G.; Walpoth, B.H.; Bowlin, G.L. A three-layered electrospun matrix to mimic native arterial architecture using polycaprolactone, elastin, and collagen: A preliminary study. *Acta Biomater.* **2010**, *6*, 2422–2433. [CrossRef] [PubMed]

47. Wise, S.; Byrom, M.; Bannon, P.; Weiss, A.; Ng, M. Electrospun Elastin-based Vascular Grafts. *Hear Lung Circ.* **2008**, *17*, S19. [CrossRef]

48. Koens, M.J.W.; Krasznai, A.G.; Hanssen, A.E.J.; Hendriks, T.; Praster, R.; Daamen, W.F.; van der Vliet, J.A.; van Kuppevelt, T.H. Vascular replacement using a layered elastin-collagen vascular graft in a porcine model: One week patency versus one month occlusion. *Organogenesis* **2015**, *11*, 105–121. [CrossRef] [PubMed]

49. Chattopadhyay, S.; Raines, R.T. Collagen-Based Biomaterials for Wound Healing. *Biopolymers* **2014**, *101*, 821–833. [CrossRef] [PubMed]

50. Shoulders, M.D.; Raines, R.T. Collagen Structure and Stability. *Annu. Rev. Biochem.* **2009**, *78*, 929–958. [CrossRef] [PubMed]

51. Tampieri, A.; Celotti, G.; Landi, E.; Sandri, M.; Roveri, N.; Falini, G. Biologically inspired synthesis of bone-like composite: Self-assembled collagen fibers/hydroxyapatite nanocrystals. *J. Biomed. Mater. Res. Part A* **2003**, *67A*, 618–625. [CrossRef] [PubMed]

52. Torculas, M.; Medina, J.; Xue, W.; Hu, X. Protein-based bioelectronics. *ACS Biomater. Sci. Eng.* **2016**, *2*, 1211–1223. [CrossRef]

53. Venugopal, J.; Low, S.; Choon, A.T.; Sampath Kumar, T.S.; Ramakrishna, S. Mineralization of osteoblasts with electrospun collagen/hydroxyapatite nanofibers. *J. Mater. Sci. Mater. Med.* **2008**, *19*, 2039–2046. [CrossRef] [PubMed]

54. Li, W.-J.; Tuli, R.; Okafor, C.; Derfoul, A.; Danielson, K.G.; Hall, D.J.; Tuan, R.S. A three-dimensional nanofibrous scaffold for cartilage tissue engineering using human mesenchymal stem cells. *Biomaterials* **2005**, *26*, 599–609. [CrossRef] [PubMed]

55. Joanne, P.; Kitsara, M.; Boitard, S.-E.; Naemetalla, H.; Vanneaux, V.; Pernot, M.; Larghero, J.; Forest, P.; Chen, Y.; Menasché, P.; et al. Nanofibrous clinical-grade collagen scaffolds seeded with human cardiomyocytes induces cardiac remodeling in dilated cardiomyopathy. *Biomaterials* **2016**, *80*, 157–168. [CrossRef] [PubMed]

56. Xu, C.Y.; Inai, R.; Kotaki, M.; Ramakrishna, S. Aligned biodegradable nanofibrous structure: A potential scaffold for blood vessel engineering. *Biomaterials* **2004**, *25*, 877–886. [CrossRef]

57. Gentleman, E.; Lay, A.N.; Dickerson, D.A.; Nauman, E.A.; Livesay, G.A.; Dee, K.C. Mechanical characterization of collagen fibers and scaffolds for tissue engineering. *Biomaterials* **2003**, *24*, 3805–3813. [CrossRef]

58. Caruso Andrea, B.; Dunn Michael, G. Functional evaluation of collagen fiber scaffolds for ACL reconstruction: Cyclic loading in proteolytic enzyme solutions. *J. Biomed. Mater. Res. Part A* **2004**, *69A*, 164–171. [CrossRef] [PubMed]

59. Kumbar, S.G.; Nukavarapu, S.P.; James, R.; Nair, L.S.; Laurencin, C.T. Electrospun poly(lactic acid-co-glycolic acid) scaffolds for skin tissue engineering. *Biomaterials* **2008**, *29*, 4100–4107. [CrossRef] [PubMed]

60. Powell, H.M.; Boyce, S.T. Engineered Human Skin Fabricated Using Electrospun Collagen–PCL Blends: Morphogenesis and Mechanical Properties. *Tissue Eng. Part A* **2009**, *15*, 2177–2187. [CrossRef] [PubMed]

61. Choi, J.S.; Lee, S.J.; Christ, G.J.; Atala, A.; Yoo, J.J. The influence of electrospun aligned poly(ε-caprolactone)/collagen nanofiber meshes on the formation of self-aligned skeletal muscle myotubes. *Biomaterials* **2008**, *29*, 2899–2906. [CrossRef] [PubMed]

62. Timnak, A.; Yousefi Gharebaghi, F.; Pajoum Shariati, R.; Bahrami, S.H.; Javadian, S.; Hojjati Emami, S.; Shokrgozar, M.A. Fabrication of nano-structured electrospun collagen scaffold intended for nerve tissue engineering. *J. Mater. Sci. Mater. Med.* **2011**, *22*, 1555–1567. [CrossRef] [PubMed]

63. Bini, T.B.; Shujun, G.; Ter Chyan, T.; Shu, W.; Aymeric, L.; Lim Ben, H.; Ramakrishna, S. Electrospun poly(L-lactide-*co*-glycolide) biodegradable polymer nanofibre tubes for peripheral nerve regeneration. *Nanotechnology* **2004**, *15*, 1459. [CrossRef]

64. Hu, Y.; Dan, W.; Xiong, S.; Kang, Y.; Dhinakar, A.; Wu, J.; Gu, Z. Development of collagen/polydopamine complexed matrix as mechanically enhanced and highly biocompatible semi-natural tissue engineering scaffold. *Acta Biomater.* **2017**, *47*, 135–148. [CrossRef] [PubMed]

65. Luo, X.; Guo, Z.; He, P.; Chen, T.; Li, L.; Ding, S.; Li, H. Study on structure, mechanical property and cell cytocompatibility of electrospun collagen nanofibers crosslinked by common agents. *Int. J. Biol. Macromol.* **2018**, *113*, 476–486. [CrossRef] [PubMed]

66. Zhang, Y.; Ouyang, H.; Lim Chwee, T.; Ramakrishna, S.; Huang, Z.M. Electrospinning of gelatin fibers and gelatin/PCL composite fibrous scaffolds. *J. Biomed. Mater. Res. Part B Appl. Biomater.* **2004**, *72B*, 156–165. [CrossRef] [PubMed]

67. Chen, T.; Embree, H.D.; Brown, E.M.; Taylor, M.M.; Payne, G.F. Enzyme-catalyzed gel formation of gelatin and chitosan: Potential for in situ applications. *Biomaterials* **2003**, *24*, 2831–2841. [CrossRef]

68. Bigi, A.; Cojazzi, G.; Panzavolta, S.; Rubini, K.; Roveri, N. Mechanical and thermal properties of gelatin films at different degrees of glutaraldehyde crosslinking. *Biomaterials* **2001**, *22*, 763–768. [CrossRef]

69. DeFrates, K.; Markiewicz, T.; Gallo, P.; Rack, A.; Weyhmiller, A.; Jarmusik, B.; Hu, X. Protein polymer-based nanoparticles: fabrication and medical applications. *Int. J. Mol. Sci.* **2018**, *19*, 1717. [CrossRef] [PubMed]

70. Rose, B.J.; Pacelli, S.; Haj, J.A.; Dua, S.H.; Hopkinson, A.; White, J.L.; Rose, R.F. Gelatin-Based Materials in Ocular Tissue Engineering. *Materials* **2014**, *7*, 3106–3135. [CrossRef] [PubMed]

71. Baradaran-Rafii, A.; Biazar, E.; Heidari-Keshel, S. Cellular Response of Limbal Stem Cells on PHBV/Gelatin Nanofibrous Scaffold for Ocular Epithelial Regeneration. *Int. J. Polym. Mater. Polym. Biomater.* **2015**, *64*, 879–887. [CrossRef]

72. Ji, W.; Yang, F.; Ma, J.; Bouma, M.J.; Boerman, O.C.; Chen, Z.; van den Beucken, J.J.J.P.; Jansen, J.A. Incorporation of stromal cell-derived factor-1α in PCL/gelatin electrospun membranes for guided bone regeneration. *Biomaterials* **2013**, *34*, 735–745. [CrossRef] [PubMed]

73. Liu, Y.; Lu, Y.; Tian, X.; Cui, G.; Zhao, Y.; Yang, Q.; Yu, S.; Xing, G.; Zhang, B. Segmental bone regeneration using an rhBMP-2-loaded gelatin/nanohydroxyapatite/fibrin scaffold in a rabbit model. *Biomaterials* **2009**, *30*, 6276–6285. [CrossRef] [PubMed]

74. Ravichandran, R.; Venugopal, J.R.; Sundarrajan, S.; Mukherjee, S.; Ramakrishna, S. Poly(Glycerol Sebacate)/Gelatin Core/Shell Fibrous Structure for Regeneration of Myocardial Infarction. *Tissue Eng. Part A* **2011**, *17*, 1363–1373. [CrossRef] [PubMed]

75. Kai, D.; Prabhakaran Molamma, P.; Jin, G.; Ramakrishna, S. Guided orientation of cardiomyocytes on electrospun aligned nanofibers for cardiac tissue engineering. *J. Biomed. Mater. Res. Part B Appl. Biomater.* **2011**, *98B*, 379–386. [CrossRef] [PubMed]

76. Ghasemi-Mobarakeh, L.; Prabhakaran, M.P.; Morshed, M.; Nasr-Esfahani, M.-H.; Ramakrishna, S. Electrospun poly(ε-caprolactone)/gelatin nanofibrous scaffolds for nerve tissue engineering. *Biomaterials* **2008**, *29*, 4532–4539. [CrossRef] [PubMed]

77. Chen, Y.-S.; Chang, J.-Y.; Cheng, C.-Y.; Tsai, F.-J.; Yao, C.-H.; Liu, B.-S. An in vivo evaluation of a biodegradable genipin-cross-linked gelatin peripheral nerve guide conduit material. *Biomaterials* **2005**, *26*, 3911–3918. [CrossRef] [PubMed]

78. Chong, E.J.; Phan, T.T.; Lim, I.J.; Zhang, Y.Z.; Bay, B.H.; Ramakrishna, S.; Lim, C.T. Evaluation of electrospun PCL/gelatin nanofibrous scaffold for wound healing and layered dermal reconstitution. *Acta Biomater.* **2007**, *3*, 321–330. [CrossRef] [PubMed]

79. Jao, D.; Mou, X.; Hu, X. Tissue Regeneration: A Silk Road. *J. Funct. Biomater.* **2016**, *7*, 22. [CrossRef] [PubMed]

80. Lefèvre, T.; Rousseau, M.-E.; Pézolet, M. Protein Secondary Structure and Orientation in Silk as Revealed by Raman Spectromicroscopy. *Biophys. J.* **2007**, *92*, 2885–2895. [CrossRef] [PubMed]

81. Vepari, C.; Kaplan, D.L. Silk as a biomaterial. *Prog. Polym. Sci.* **2007**, *32*, 991–1007. [CrossRef] [PubMed]

82. Najjar, R.; Luo, Y.; Jao, D.; Brennan, D.; Xue, Y.; Beachley, V.; Hu, X.; Xue, W. Biocompatible silk/polymer energy harvesters using stretched poly (vinylidene fluoride-*co*-hexafluoropropylene) (PVDF-HFP) nanofibers. *Polymers* **2017**, *9*, 479. [CrossRef]

83. Meinel, L.; Karageorgiou, V.; Hofmann, S.; Fajardo, R.; Snyder, B.; Li, C.; Zichner, L.; Langer, R.; Vunjak-Novakovic, G.; Kaplan, D.L. Engineering bone-like tissue in vitro using human bone marrow stem cells and silk scaffolds. *J. Biomed. Mater. Res. Part A* **2004**, *71A*, 25–34. [CrossRef] [PubMed]

84. Sommer, M.R.; Vetsch, J.R.; Leemann, J.; Müller, R.; Studart, A.R.; Hofmann, S. Silk fibroin scaffolds with inverse opal structure for bone tissue engineering. *J. Biomed. Mater. Res. Part B Appl. Biomater.* **2017**, *105*, 2074–2084. [CrossRef] [PubMed]

85. Aliramaji, S.; Zamanian, A.; Mozafari, M. Super-paramagnetic responsive silk fibroin/chitosan/magnetite scaffolds with tunable pore structures for bone tissue engineering applications. *Mater. Sci. Eng. C* **2017**, *70*, 736–744. [CrossRef] [PubMed]

86. Vishwanath, V.; Pramanik, K.; Biswas, A. Development of a novel glucosamine/silk fibroin–chitosan blend porous scaffold for cartilage tissue engineering applications. *Iran. Polym. J.* **2017**, *26*, 11–19. [CrossRef]

87. Du, J.; Zhu, T.; Yu, H.; Zhu, J.; Sun, C.; Wang, J.; Chen, S.; Wang, J.; Guo, X. Potential applications of three-dimensional structure of silk fibroin/poly(ester-urethane) urea nanofibrous scaffold in heart valve tissue engineering. *Appl. Surf. Sci.* **2018**, *447*, 269–278. [CrossRef]

88. Rao, J.; Cheng, Y.; Liu, Y.; Ye, Z.; Zhan, B.; Quan, D.; Xu, Y. A multi-walled silk fibroin/silk sericin nerve conduit coated with poly(lactic-*co*-glycolic acid) sheath for peripheral nerve regeneration. *Mater. Sci. Eng. C* **2017**, *73*, 319–332. [CrossRef] [PubMed]

89. Zhang, W.; Chen, L.; Chen, J.; Wang, L.; Gui, X.; Ran, J.; Xu, G.; Zhao, H.; Zeng, M.; Ji, J.; et al. Silk Fibroin Biomaterial Shows Safe and Effective Wound Healing in Animal Models and a Randomized Controlled Clinical Trial. *Adv. Healthc. Mater.* **2017**, *6*, 1700121. [CrossRef] [PubMed]

90. Min, B.-M.; Lee, G.; Kim, S.H.; Nam, Y.S.; Lee, T.S.; Park, W.H. Electrospinning of silk fibroin nanofibers and its effect on the adhesion and spreading of normal human keratinocytes and fibroblasts in vitro. *Biomaterials* **2004**, *25*, 1289–1297. [CrossRef] [PubMed]

91. Hafner, K.; Montag, D.; Maeser, H.; Peng, C.; Marcotte, W.R.; Dean, D.; Kennedy, M.S. Evaluating adhesion and alignment of dental pulp stem cells to a spider silk substrate for tissue engineering applications. *Mater. Sci. Eng. C* **2017**, *81*, 104–112. [CrossRef] [PubMed]

92. Hou, J.; Xie, Y.; Ji, A.; Cao, A.; Fang, Y.; Shi, E. Carbon-Nanotube-Wrapped Spider Silks for Directed Cardiomyocyte Growth and Electrophysiological Detection. *ACS Appl. Mater. Interfaces* **2018**, *10*, 6793–6798. [CrossRef] [PubMed]

93. Rouse, J.G.; Van Dyke, M.E. A Review of Keratin-Based Biomaterials for Biomedical Applications. *Materials* **2010**, *3*, 999–1014. [CrossRef]

94. MacKinnon, P.J.; Powell, B.C.; Rogers, G.E. Structure and expression of genes for a class of cysteine-rich proteins of the cuticle layers of differentiating wool and hair follicles. *J. Cell Biol.* **1990**, *111*, 2587–2600. [CrossRef] [PubMed]

95. Zhu, H.; Li, R.; Wu, X.; Chen, K.; Che, J. Controllable fabrication and characterization of hydrophilic PCL/wool keratin nanonets by electronetting. *Eur. Polym. J.* **2017**, *86*, 154–161. [CrossRef]

96. Esparza, Y.; Ullah, A.; Boluk, Y.; Wu, J. Preparation and characterization of thermally crosslinked poly(vinyl alcohol)/feather keratin nanofiber scaffolds. *Mater. Des.* **2017**, *133*, 1–9. [CrossRef]

97. Arslan, Y.E.; Sezgin Arslan, T.; Derkus, B.; Emregul, E.; Emregul, K.C. Fabrication of human hair keratin/jellyfish collagen/eggshell-derived hydroxyapatite osteoinductive biocomposite scaffolds for bone tissue engineering: From waste to regenerative medicine products. *Colloids Surf. B Biointerfaces* **2017**, *154*, 160–170. [CrossRef] [PubMed]

98. Xu, S.; Sang, L.; Zhang, Y.; Wang, X.; Li, X. Biological evaluation of human hair keratin scaffolds for skin wound repair and regeneration. *Mater. Sci. Eng. C* **2013**, *33*, 648–655. [CrossRef] [PubMed]

99. Xu, H.; Cai, S.; Xu, L.; Yang, Y. Water-Stable Three-Dimensional Ultrafine Fibrous Scaffolds from Keratin for Cartilage Tissue Engineering. *Langmuir* **2014**, *30*, 8461–8470. [CrossRef] [PubMed]

100. Hualin, Z.; Jinsong, L. Electrospun poly(lactic-*co*-glycolic acid)/wool keratin fibrous composite scaffolds potential for bone tissue engineering applications. *J. Bioact. Compat. Polym.* **2013**, *28*, 141–153.

101. Wang, J.Y.; Fujimoto, K.; Miyazawa, T.; Endo, Y. Antioxidative mechanism of maize zein in powder model systems against methyl linoleate: Effect of water activity and coexistence of antioxidants. *J. Agric. Food Chem.* **1991**, *39*, 351–355. [CrossRef]

102. Dong, J.; Sun, Q.; Wang, J.-Y. Basic study of corn protein, zein, as a biomaterial in tissue engineering, surface morphology and biocompatibility. *Biomaterials* **2004**, *25*, 4691–4697. [CrossRef] [PubMed]

103. Wang, H.-J.; Di, L.; Ren, Q.-S.; Wang, J.-Y. Applications and Degradation of Proteins Used as Tissue Engineering Materials. *Materials* **2009**, *2*, 613–635. [CrossRef]

104. Demir, M.; Ramos-Rivera, L.; Silva, R.; Nazhat, S.N.; Boccaccini, A.R. Zein-based composites in biomedical applications. *J. Biomed. Mater. Res. Part A* **2017**, *105*, 1656–1665. [CrossRef] [PubMed]

105. Yang, F.; Miao, Y.; Wang, Y.; Zhang, L.-M.; Lin, X. Electrospun Zein/Gelatin Scaffold-Enhanced Cell Attachment and Growth of Human Periodontal Ligament Stem Cells. *Materials* **2017**, *10*, 1168. [CrossRef] [PubMed]
106. Vogt, L.; Liverani, L.; Roether, J.; Boccaccini, A. Electrospun Zein Fibers Incorporating Poly(glycerol sebacate) for Soft Tissue Engineering. *Nanomaterials* **2018**, *8*, 150. [CrossRef] [PubMed]
107. Jiang, Q.; Reddy, N.; Yang, Y. Cytocompatible cross-linking of electrospun zein fibers for the development of water-stable tissue engineering scaffolds. *Acta Biomater.* **2010**, *6*, 4042–4051. [CrossRef] [PubMed]
108. Paliwal, R.; Palakurthi, S. Zein in controlled drug delivery and tissue engineering. *J. Control. Release Off. J. Control. Release Soc.* **2014**, *189*, 108–122. [CrossRef] [PubMed]
109. Bouman, J.; Belton, P.; Venema, P.; van der Linden, E.; de Vries, R.; Qi, S. Controlled Release from Zein Matrices: Interplay of Drug Hydrophobicity and pH. *Pharm. Res.* **2016**, *33*, 673–685. [CrossRef] [PubMed]
110. Labib, G. Overview on zein protein: A promising pharmaceutical excipient in drug delivery systems and tissue engineering. *Expert Opin. Drug Deliv.* **2018**, *15*, 65–75. [CrossRef] [PubMed]
111. Tansaz, S.; Boccaccini, A.R. Biomedical applications of soy protein: A brief overview. *J. Biomed. Mater. Res. Part A* **2016**, *104*, 553–569. [CrossRef] [PubMed]
112. Li, S.; Donner, E.; Xiao, H.; Thompson, M.; Zhang, Y.; Rempel, C.; Liu, Q. Preparation and characterization of soy protein films with a durable water resistance-adjustable and antimicrobial surface. *Mater. Sci. Eng. C* **2016**, *69*, 947–955. [CrossRef] [PubMed]
113. Maruthi, Y.; Sudhakar, H.; Rao, U.S.; Babu, P.K.; Rao, K.C.; Subha, M.C.S. Blend Membranes of Sodium alginate and Soya protein for Pervaporation Dehydration of Isopropanol. *Adv. Polym. Sci. Technol.* **2014**, *4*, 12–21.
114. Tansaz, S.; Liverani, L.; Vester, L.; Boccaccini, A.R. Soy protein meets bioactive glass: Electrospun composite fibers for tissue engineering applications. *Mater. Lett.* **2017**, *199*, 143–146. [CrossRef]
115. Ahn, S.; Chantre, C.O.; Gannon, A.R.; Lind, J.U.; Campbell, P.H.; Grevesse, T.; O'Connor, B.B.; Parker, K.K. Soy Protein/Cellulose Nanofiber Scaffolds Mimicking Skin Extracellular Matrix for Enhanced Wound Healing. *Adv. Healthc. Mater.* **2018**, *7*, 1701175. [CrossRef] [PubMed]
116. Xu, H.; Cai, S.; Sellers, A.; Yang, Y. Intrinsically water-stable electrospun three-dimensional ultrafine fibrous soy protein scaffolds for soft tissue engineering using adipose derived mesenchymal stem cells. *RSC Adv.* **2014**, *4*, 15451–15457. [CrossRef]
117. Silva, S.S.; Goodfellow, B.J.; Benesch, J.; Rocha, J.; Mano, J.F.; Reis, R.L. Morphology and miscibility of chitosan/soy protein blended membranes. *Carbohydr. Polym.* **2007**, *70*, 25–31. [CrossRef]
118. Beachley, V.; Wen, X. Effect of electrospinning parameters on the nanofiber diameter and length. *Mater. Sci. Eng. C Mater. Biol. Appl.* **2009**, *29*, 663–668. [CrossRef] [PubMed]
119. Charernsriwilaiwat, N.; Opanasopit, P.; Rojanarata, T.; Ngawhirunpat, T.; Supaphol, P. Preparation and characterization of chitosan-hydroxybenzotriazole/polyvinyl alcohol blend nanofibers by the electrospinning technique. *Carbohydr. Polym.* **2010**, *81*, 675–680. [CrossRef]
120. Pakravan, M.; Heuzey, M.-C.; Ajji, A. A fundamental study of chitosan/PEO electrospinning. *Polymer* **2011**, *52*, 4813–4824. [CrossRef]
121. Tan, S.H.; Inai, R.; Kotaki, M.; Ramakrishna, S. Systematic parameter study for ultra-fine fiber fabrication via electrospinning process. *Polymer* **2005**, *46*, 6128–6134. [CrossRef]
122. Arafat, M.T.; Tronci, G.; Yin, J.; Wood, D.J.; Russell, S.J. Biomimetic wet-stable fibres via wet spinning and diacid-based crosslinking of collagen triple helices. *Polymer* **2015**, *77*, 102–112. [CrossRef]
123. Yan, J.; Zhou, G.; Knight, D.P.; Shao, Z.; Chen, X. Wet-Spinning of Regenerated Silk Fiber from Aqueous Silk Fibroin Solution: Discussion of Spinning Parameters. *Biomacromolecules* **2010**, *11*, 1–5. [CrossRef] [PubMed]
124. Plaza, G.R.; Corsini, P.; Marsano, E.; Pérez-Riguiero, J.; Elices, M.; Riekel, C.; Vendrely, C.; Guinea, G.V. Correlation between processing conditions, microstructure and mechanical behavior in regenerated silkworm silk fibers. *J. Polym. Sci. Part B Polym. Phys.* **2012**, *50*, 455–465. [CrossRef]
125. Um, I.C.; Ki, C.S.; Kweon, H.; Lee, K.G.; Ihm, D.W.; Park, Y.H. Wet spinning of silk polymer. II. Effect of drawing on the structural characteristics and properties of filament. *Int. J. Biol. Macromol.* **2004**, *34*, 107–119. [CrossRef] [PubMed]
126. Wei, W.; Zhang, Y.; Zhao, Y.; Luo, J.; Shao, H.; Hu, X. Bio-inspired capillary dry spinning of regenerated silk fibroin aqueous solution. *Mater. Sci. Eng. C* **2011**, *31*, 1602–1608. [CrossRef]

127. Xie, J.; Wang, C.H. Electrospun micro and nanofibers for sustained delivery of paclitaxel to treat C6 glioma in vitro. *Pharm. Res.* **2006**, *23*, 1817. [CrossRef] [PubMed]

128. Loordhuswamy, A.M.; Krishnaswamy, V.R.; Korrapati, P.S.; Thinakaran, S.; Rengaswami, G.D. Fabrication of highly aligned fibrous scaffolds for tissue regeneration by centrifugal spinning technology. *Mater. Sci. Eng. C Mater. Biol. Appl.* **2014**, *42*, 799–807. [CrossRef] [PubMed]

129. Zhang, X.; Lu, Y. Centrifugal Spinning: An Alternative Approach to Fabricate Nanofibers at High Speed and Low Cost. *Polym. Rev.* **2014**, *54*, 677–701. [CrossRef]

130. Medeiros, E.S.; Glenn, G.M.; Klamczynski, A.P.; Orts, W.J.; Mattoso, L.H.C. Solution blow spinning: A new method to produce micro- and nanofibers from polymer solutions. *J. Appl. Polym. Sci.* **2009**, *113*, 2322–2330. [CrossRef]

131. Gupta, K.C.; Haider, A.; Choi, Y.-R.; Kang, I.-K. Nanofibrous scaffolds in biomedical applications. *Biomater. Res.* **2014**, *18*, 5. [CrossRef] [PubMed]

132. Matthews, J.A.; Wnek, G.E.; Simpson, D.G.; Bowlin, G.L. Electrospinning of Collagen Nanofibers. *Biomacromolecules* **2002**, *3*, 232–238. [CrossRef] [PubMed]

133. Subramanian, A.; Krishnan, U.M.; Sethuraman, S. Fabrication of uniaxially aligned 3D electrospun scaffolds for neural regeneration. *Biomed. Mater.* **2011**, *6*, 025004. [CrossRef] [PubMed]

134. Goh, Y.F.; Shakir, I.; Hussain, R. Electrospun fibers for tissue engineering, drug delivery, and wound dressing. *J. Mater. Sci.* **2013**, *48*, 3027–3054. [CrossRef]

135. Zahedi, P.; Rezaeian, I.; Ranaei-Siadat, S.-O.; Jafari, S.-H.; Supaphol, P. A review on wound dressings with an emphasis on electrospun nanofibrous polymeric bandages. *Polym. Adv. Technol.* **2010**, *21*, 77–95. [CrossRef]

136. Wharram, S.E.; Zhang, X.; Kaplan, D.L.; McCarthy, S.P. Electrospun silk material systems for wound healing. *Macromol. Biosci.* **2010**, *10*, 246–257. [CrossRef] [PubMed]

137. Khadka, D.B.; Haynie, D.T. Protein- and peptide-based electrospun nanofibers in medical biomaterials. *Nanomedicine* **2012**, *8*, 1242–1262. [CrossRef] [PubMed]

138. Taylor, G.I. Electrically driven jets. *Proc. R. Soc. Lond. Ser. A* **1969**, *313*, 453–475. [CrossRef]

139. Rho, K.S.; Jeong, L.; Lee, G.; Seo, B.-M.; Park, Y.J.; Hong, S.-D.; Roh, S.; Cho, J.J.; Park, W.H.; Min, B.-M. Electrospinning of collagen nanofibers: Effects on the behavior of normal human keratinocytes and early-stage wound healing. *Biomaterials* **2006**, *27*, 1452–1461. [CrossRef] [PubMed]

140. Pegg, C.E.; Jones, G.H.; Athauda, T.J.; Ozer, R.R.; Chalker, J.M. Facile preparation of ammonium alginate-derived nanofibers carrying diverse therapeutic cargo. *Chem. Commun.* **2014**, *50*, 156–158. [CrossRef] [PubMed]

141. Li, G.; Li, Y.; Chen, G.; He, J.; Han, Y.; Wang, X.; Kaplan, D.L. Silk-based biomaterials in biomedical textiles and fiber-based implants. *Adv. Healthc. Mater.* **2015**, *4*, 1134–1151. [CrossRef] [PubMed]

142. Lee, K.H.; Baek, D.H.; Ki, C.S.; Park, Y.H. Preparation and characterization of wet spun silk fibroin/poly(vinyl alcohol) blend filaments. *Int. J. Biol. Macromol.* **2007**, *41*, 168–172. [CrossRef] [PubMed]

143. Qiu, W.; Teng, W.; Cappello, J.; Wu, X. Wet-Spinning of Recombinant Silk-Elastin-Like Protein Polymer Fibers with High Tensile Strength and High Deformability. *Biomacromolecules* **2009**, *10*, 602–608. [CrossRef] [PubMed]

144. Sun, M.; Zhang, Y.; Zhao, Y.; Shao, H.; Hu, X. The structure–property relationships of artificial silk fabricated by dry-spinning process. *J. Mater. Chem.* **2012**, *22*, 18372–18379. [CrossRef]

145. Tian, Q.; Xu, Z.; Liu, Y.; Fang, B.; Peng, L.; Xi, J.; Li, Z.; Gao, C. Dry spinning approach to continuous graphene fibers with high toughness. *Nanoscale* **2017**, *9*, 12335–12342. [CrossRef] [PubMed]

146. Nain, A.; Wong, J.; Amon, C.; Sitti, M. Drawing suspended polymer micro-/nanofibers using glass micropipettes. *Appl. Phys. Lett.* **2006**, *89*, 183105. [CrossRef]

147. Buzgo, M.; Rampichova, M.; Vocetkova, K.; Sovkova, V.; Lukasova, V.; Doupnik, M.; Mickova, A.; Rustichelli, F.; Amler, E. Emulsion centrifugal spinning for production of 3D drug releasing nanofibres with core/shell structure. *RSC Adv.* **2017**, *7*, 1215–1228. [CrossRef]

148. Weitz, R.T.; Harnau, L.; Rauschenbach, S.; Burghard, M.; Kern, K. Polymer Nanofibers via Nozzle-Free Centrifugal Spinning. *Nano Lett.* **2008**, *8*, 1187–1191. [CrossRef] [PubMed]

149. Metera, A.; Wojasiński, M.; Ciach, T. Production of Polymer Fibres by Solution Blow Spinning. In Proceedings of the 2nd European Young Engineers Conference 2013, Newcastle upon Tyne, UK, 4 July 2013.

150. Mandal, D.; Nasrolahi Shirazi, A.; Parang, K. Self-assembly of peptides to nanostructures. *Org. Biomol. Chem.* **2014**, *12*, 3544–3561. [CrossRef] [PubMed]

151. Zhang, S. Fabrication of novel biomaterials through molecular self-assembly. *Nat. Biotechnol.* **2003**, *21*, 1171–1178. [CrossRef] [PubMed]
152. Nayak, R.; Padhye, R.; Kyratzis, I.L.; Truong, Y.B.; Arnold, L. Recent advances in nanofibre fabrication techniques. *Text. Res. J.* **2011**, *82*, 129–147. [CrossRef]
153. Hwang, W.; Kim, B.H.; Dandu, R.; Cappello, J.; Ghandehari, H.; Seog, J. Surface Induced nanofiber growth by self-assembly of a silk-elastin-like protein polymer. *Langmuir* **2009**, *25*, 12682–12686. [CrossRef] [PubMed]
154. Beachley, V.; Wen, X. Polymer nanofibrous structures: Fabrication, biofunctionalization, and cell interactions. *Prog. Polym. Sci.* **2010**, *35*, 868–892. [CrossRef] [PubMed]
155. Bajáková, J.; Chaloupek, J.; Lukáš, D.; Lacarin, M. Drawing—The production of individual nanofibers by experimental method. In Proceedings of the 3rd International Conference on Nanotechnology-Smart Materials (NANOCON'11), Brno, Czech Republic, 21–23 September 2011.
156. Ondarçuhu, T.; Joachim, C. Drawing a single nanofibre over hundreds of microns. *EPL* **1998**, *42*, 215. [CrossRef]
157. Feughelman, M. Natural protein fibers. *J. Appl. Polym. Sci.* **2002**, *83*, 489–507. [CrossRef]
158. Bonassar, L.J.; Vacanti, C.A. Tissue engineering: the first decade and beyond. *J. Cell. Biochem. Suppl.* **1998**, *72*, 297–303. [CrossRef]
159. Celikkin, N.; Rinoldi, C.; Costantini, M.; Trombetta, M.; Rainer, A.; Swieszkowski, W. Naturally derived proteins and glycosaminoglycan scaffolds for tissue engineering applications. *Mater. Sci. Eng. C* **2017**, *78*, 1277–1299. [CrossRef] [PubMed]
160. López-Noriega, A.; Quinlan, E.; Celikkin, N.; O'Brien, F.J. Incorporation of polymeric microparticles into collagen-hydroxyapatite scaffolds for the delivery of a pro-osteogenic peptide for bone tissue engineering. *APL Mater.* **2015**, *3*, 014910. [CrossRef]
161. Browning, M.; Dempsey, D.; Guiza, V.; Becerra, S.; Rivera, J.; Russell, B.; Höök, M.; Clubb, F.; Miller, M.; Fossum, T. Multilayer vascular grafts based on collagen-mimetic proteins. *Acta Biomater.* **2012**, *8*, 1010–1021. [CrossRef] [PubMed]
162. Farrell, E.; O'Brien, F.J.; Doyle, P.; Fischer, J.; Yannas, I.; Harley, B.A.; O'Connell, B.; Prendergast, P.J.; Campbell, V.A. A collagen-glycosaminoglycan scaffold supports adult rat mesenchymal stem cell differentiation along osteogenic and chondrogenic routes. *Tissue Eng.* **2006**, *12*, 459–468. [CrossRef] [PubMed]
163. Liu, Y.; Ma, L.; Gao, C. Facile fabrication of the glutaraldehyde cross-linked collagen/chitosan porous scaffold for skin tissue engineering. *Mater. Sci. Eng. C* **2012**, *32*, 2361–2366. [CrossRef]
164. Calderón-Colón, X.; Xia, Z.; Breidenich, J.L.; Mulreany, D.G.; Guo, Q.; Uy, O.M.; Tiffany, J.E.; Freund, D.E.; McCally, R.L.; Schein, O.D. Structure and properties of collagen vitrigel membranes for ocular repair and regeneration applications. *Biomaterials* **2012**, *33*, 8286–8295. [CrossRef] [PubMed]
165. Soller, E.C.; Tzeranis, D.S.; Miu, K.; So, P.T.; Yannas, I.V. Common features of optimal collagen scaffolds that disrupt wound contraction and enhance regeneration both in peripheral nerves and in skin. *Biomaterials* **2012**, *33*, 4783–4791. [CrossRef] [PubMed]
166. Saxena, T.; Karumbaiah, L.; Valmikinathan, C.M. *Proteins and Poly (Amino Acids)*; Elsevier: New York, NY, USA, 2014; Chapter 3.
167. Ribeiro, N.; Sousa, S.R.; van Blitterswijk, C.A.; Moroni, L.; Monteiro, F.J. A biocomposite of collagen nanofibers and nanohydroxyapatite for bone regeneration. *Biofabrication* **2014**, *6*, 035015. [CrossRef] [PubMed]
168. Tillman, B.W.; Yazdani, S.K.; Lee, S.J.; Geary, R.L.; Atala, A.; Yoo, J.J. The in vivo stability of electrospun polycaprolactone-collagen scaffolds in vascular reconstruction. *Biomaterials* **2009**, *30*, 583–588. [CrossRef] [PubMed]
169. Ekaputra, A.K.; Prestwich, G.D.; Cool, S.M.; Hutmacher, D.W. The three-dimensional vascularization of growth factor-releasing hybrid scaffold of poly (epsilon-caprolactone)/collagen fibers and hyaluronic acid hydrogel. *Biomaterials* **2011**, *32*, 8108–8117. [CrossRef] [PubMed]
170. Melke, J.; Midha, S.; Ghosh, S.; Ito, K.; Hofmann, S. Silk fibroin as biomaterial for bone tissue engineering. *Acta Biomater.* **2016**, *31*, 1–16. [CrossRef] [PubMed]
171. Kim, K.H.; Jeong, L.; Park, H.N.; Shin, S.Y.; Park, W.H.; Lee, S.C.; Kim, T.I.; Park, Y.J.; Seol, Y.J.; Lee, Y.M.; et al. Biological efficacy of silk fibroin nanofiber membranes for guided bone regeneration. *J. Biotechnol.* **2005**, *120*, 327–339. [CrossRef] [PubMed]

172. Jao, D.; Xue, Y.; Medina, J.; Hu, X. Protein-based drug-delivery materials. *Materials* **2017**, *10*, 517. [CrossRef] [PubMed]

173. Huang, W.; Zou, T.; Li, S.; Jing, J.; Xia, X.; Liu, X. Drug-loaded zein nanofibers prepared using a modified coaxial electrospinning process. *AAPS PharmSciTech* **2013**, *14*, 675–681. [CrossRef] [PubMed]

174. Xu, W.; Yang, Y. Drug sorption onto and release from soy protein fibers. *J. Mater. Sci. Mater. Med.* **2009**, *20*, 2477–2486. [CrossRef] [PubMed]

175. Lee, F.-Y.; Lee, D.; Lee, T.-C.; Chen, J.-K.; Wu, R.-C.; Liu, K.-C.; Liu, S.-J. Fabrication of Multi-Layered Lidocaine and Epinephrine-Eluting PLGA/Collagen Nanofibers: In Vitro and In Vivo Study. *Polymers* **2017**, *9*, 416. [CrossRef]

176. Zeng, J.; Aigner, A.; Czubayko, F.; Kissel, T.; Wendorff, J.H.; Greiner, A. Poly(vinyl alcohol) Nanofibers by Electrospinning as a Protein Delivery System and the Retardation of Enzyme Release by Additional Polymer Coatings. *Biomacromolecules* **2005**, *6*, 1484–1488. [CrossRef] [PubMed]

177. Boateng, J.S.; Matthews, K.H.; Stevens, H.N.E.; Eccleston, G.M. Wound healing dressings and drug delivery systems: A review. *J. Pharm. Sci.* **2008**, *97*, 2892–2923. [CrossRef] [PubMed]

178. Rieger, K.A.; Birch, N.P.; Schiffman, J.D. Designing electrospun nanofiber mats to promote wound healing—A review. *J. Mater. Chem. B* **2013**, *1*, 4531–4541. [CrossRef]

179. Wang, Y.; Li, P.; Xiang, P.; Lu, J.; Yuan, J.; Shen, J. Electrospun polyurethane/keratin/AgNP biocomposite mats for biocompatible and antibacterial wound dressings. *J. Mater. Chem. B* **2016**, *4*, 635–648. [CrossRef]

180. Yuan, J.; Xing, Z.-C.; Park, S.-W.; Geng, J.; Kang, I.-K.; Yuan, J.; Shen, J.; Meng, W.; Shim, K.-J.; Han, I.-S.; et al. Fabrication of PHBV/keratin composite nanofibrous mats for biomedical applications. *Macromol. Res.* **2009**, *17*, 850–855. [CrossRef]

nanomaterials

MDPI

Review

Graphene-Based Nanomaterials for Tissue Engineering in the Dental Field

Riccardo Guazzo [1,†] , Chiara Gardin [2,3,†], Gloria Bellin [2,3] , Luca Sbricoli [1] ,
Letizia Ferroni [2,3,*], Francesco Saverio Ludovichetti [1] , Adriano Piattelli [4],
Iulian Antoniac [5] , Eriberto Bressan [1] and Barbara Zavan [2,6,*]

[1]   Department of Neurosciences, Institute of Clinical Dentistry, University of Padova, 35128 Padova, Italy;
       riccardo.guazzo@unipd.it (R.G.); luca.sbricoli@unipd.it (L.S.); francesco.ludovichetti@unipd.it (F.S.L.);
       eriberto.bressan@unipd.it (E.B.)
[2]   Department of Biomedical Sciences, University of Padova, 35131 Padova, Italy;
       chiara.gardin@unipd.it (C.G.); gloria.bellin@gmail.com (G.B.)
[3]   Maria Pia Hospital, GVM Care & Research, 10132 Torino, Italy
[4]   Department of Medical, Oral and Biotechnological Sciences, University of Chieti-Pescara,
       66100 Chieti, Italy; apiattelli@unich.it
[5]   Department Materials Science and Engineering, University Politehnica of Bucharest,
       060032 Bucharest, Romania; antoniac.iulian@gmail.com
[6]   Maria Cecilia Hospital, GVM Care & Research, 48033 Ravenna, Italy
*   Correspondence: letizia.ferroni@unipd.it (L.F.); barbara.zavan@unipd.it (B.Z.); Tel.: +39-049-827-6096 (B.Z.)
†   These authors contributed equally to this work.

Received: 3 May 2018; Accepted: 17 May 2018; Published: 20 May 2018

check for
updates

Abstract: The world of dentistry is approaching graphene-based nanomaterials as substitutes for tissue engineering. Apart from its exceptional mechanical strength, electrical conductivity and thermal stability, graphene and its derivatives can be functionalized with several bioactive molecules. They can also be incorporated into different scaffolds used in regenerative dentistry, generating nanocomposites with improved characteristics. This review presents the state of the art of graphene-based nanomaterial applications in the dental field. We first discuss the interactions between cells and graphene, summarizing the available in vitro and in vivo studies concerning graphene biocompatibility and cytotoxicity. We then highlight the role of graphene-based nanomaterials in stem cell control, in terms of adhesion, proliferation and differentiation. Particular attention will be given to stem cells of dental origin, such as those isolated from dental pulp, periodontal ligament or dental follicle. The review then discusses the interactions between graphene-based nanomaterials with cells of the immune system; we also focus on the antibacterial activity of graphene nanomaterials. In the last section, we offer our perspectives on the various opportunities facing the use of graphene and its derivatives in associations with titanium dental implants, membranes for bone regeneration, resins, cements and adhesives as well as for tooth-whitening procedures.

Keywords: graphene; nanomaterials; dental stem cells; antibacterial activity; dental implant; bone regeneration

1. Introduction

Tissue engineering is an interdisciplinary science which aims at developing biological substitutes to restore, maintain, or improve tissue function by using a combination of cells, scaffolds and suitable biochemical factors [1]. Scaffolds in particular represent the key element in tissue engineering research, whose role is not only to provide the appropriate environment for specific cells but also to retain growth and nutrition factors for cellular migration, adhesion, growth and differentiation [2].

Over the years, several natural and synthetic three-dimensional (3D) scaffolds have been successfully developed and employed for various tissues, such as skin [3], cartilage [4], muscle [5], vasculature [6] and bone [7]. With regard to dental tissues, regeneration is very challenging and requires thorough understanding of biological events at the cellular and molecular level [8]. Tooth indeed is one of the most difficult tissues to treat due to its heterogeneous and dynamic anatomical structure which includes the vital dentin-pulp complex, cementum, periodontal ligament, alveolar bone and enamel [9]. Furthermore, dental tissues show a limited and variable degree for self-repair as a result of injury or disease: cementum, for example, has a very slow regenerative capacity, whereas enamel regeneration is not possible. Dentin can regenerate, while dental pulp has a partial regeneration capacity as it is enclosed in dentin and has limited apical blood supply. In contrast, alveolar bone exhibits rapid turnover in response to mechanical stimulation [10].

In the last decade, many advances have been made in the dental tissue engineering field aimed at the regeneration of dental pulp [11], periodontal ligament [12,13], dentin [14], enamel [15] and integrated tooth tissues [16,17], which have seen the interplay of stem cells, growth factors and scaffolds. Scaffolds including collagen [18], polymers [19], self-assembling peptides [20], or silk [21] have been used for dental tissues regeneration. A novel strategy that may improve the success of scaffold therapy is represented by nanosized materials [22]. Nanomaterials possess exciting physicochemical and biological properties for biomedical applications due to their small size, large surface area and ability to interact with cells, promoting their adhesion, migration, proliferation and differentiation [23].

In the last few years, graphene and its derivatives have emerged as a new class of nanomaterials [24]. Graphene is a carbon-based flat monolayer, arranged in a two-dimensional hexagonal structure, with unique mechanical, electrochemical and physical properties. The graphene family nanomaterials include several graphene derivatives, such as Few-Layered Graphene (FLG), ultrathin graphite, Graphene Oxide (GO), reduced Graphene Oxide (rGO) and graphene nanosheets [25]. These differ from each other in terms of surface properties, number of layers and size. In addition to the above graphene derivatives, graphene family also comprises graphene-based composites, which derive from functionalization of graphene with polymers, small molecules, or nanoparticles through covalent or noncovalent interactions [26] (Figure 1). Graphene surface functionalization with molecules of diverse nature allows the development of different devices, that can enhance or alter the properties required for specific application.

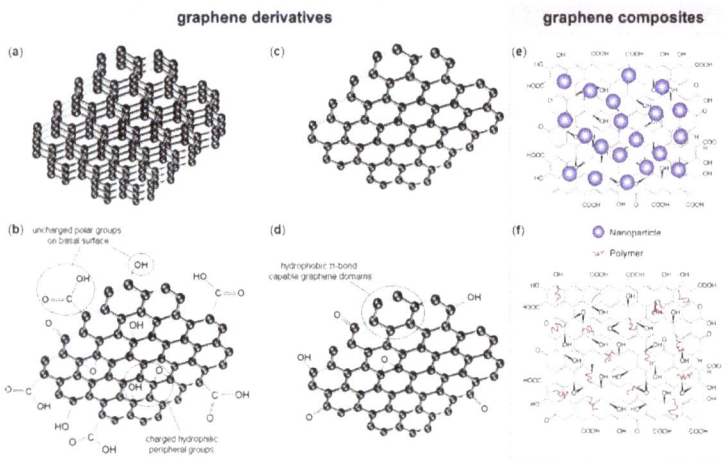

Figure 1. Schematic representation of different graphene-based nanomaterials. (**a**) Few-Layered Graphene (FLG), (**b**) Graphene Oxide (GO), (**c**) graphene nanosheets and (**d**) reduced Graphene Oxide (rGO) belong to the graphene derivatives group; (**e**) GO nanoparticle composite and (**f**) GO polymer composite are composites of graphene. Reproduced with permissions from [25,26].

Since its discovery in 2004 [27], many applications have been explored for graphene, ranging from electronic and optoelectronic devices to photoconductive materials. However, only in 2008 graphene was introduced for the first time in the field of biomedical sciences [28] and widely used in biomedical applications such as bioelectronics, bioimaging, drug delivery and tissue engineering [24].

The present review intends to provide the reader an overview of the current state of the art of the graphene-based nanomaterials in tissue engineering in dentistry. Different aspects of the graphene-based nanomaterials will be discussed in detail. First, we highlight the role of stem cells and their interactions with graphene nanomaterials. We consider the influence of graphene-based nanomaterials both on biocompatibility, cytotoxicity and differentiation properties of stem cells. In this context, particular attention will be given to stem cells of dental origin. The interactions with immune cells and antibacterial properties of graphene, which are critical aspects to consider in the hostile environment of the oral cavity, are subsequently discussed. Finally, applications of graphene-based nanomaterials on dental implants, membranes, resins, cements and adhesives as well as on teeth-whitening are extensively presented.

2. Graphene-Based Nanomaterials and Interactions with Cells

The first aspect to consider when developing a new nanomaterial for biomedical applications is its biocompatibility. Excellent biocompatibility is indeed essential in order to avoid any adverse effect of a material in living tissues [29]. Apart from being biocompatible, scaffolds should direct the successful transformation of stem cells into cells and tissues with morphological features and physiological functions similar to those in vivo. In other words, tissue engineering techniques can be considered efficient when they obtain stem cell differentiation into the desired tissue lineage [30]. Furthermore, an ideal scaffold should not evoke an inflammatory response and desirably inhibit bacterial growth on the surfaces. This feature is very critical when considering the environment of the oral cavity, which is known to be an important site for bacterial biofilm formation. In the following paragraphs, the effects of incorporation of graphene-based nanomaterials into different scaffolds, in relation to the issues described above, will be discussed in detail.

2.1. Graphene Biocompatibility In Vitro and In Vivo

Graphene-based nanomaterials are rapidly spreading as scaffolds in tissue engineering applications, also in the dental field; consequently, assessments of their biocompatibility and cytotoxicity have to be done. Dosages, viability assays, type of nanomaterials, type of cells and uses need to be evaluated before developing a new reliable graphene-based nanomaterial.

Over the years, biocompatibility of graphene and its derivatives has been largely discussed among authors. The comprehension of the toxicological potential of graphene-based nanomaterials, indeed, has to be carefully evaluated in order not to interrupt future in vivo studies and translational efforts [31]. Accumulating evidence suggests that the cytotoxicity of graphene and its derivatives is influenced by numerous factors, such as their concentration, shape, size, dispersibility, surface functionalization and that the common mechanisms describing the cytotoxicity of graphene-based nanomaterials include Reactive Oxygen Species (ROS) production and cell membrane damage [32,33].

The 3-(4,5-Dimethylthiazol-2-yl)-2, 5-Diphenyltetrazolium bromide (MTT) assay, which measures the mitochondria function of the cells, is one of the preferred method for evaluating the effect of nanomaterials in cell culture [34]. In one of the pioneering works, the in vitro toxicity of FLG sheets was compared to that of Single Wall Carbon Nanotubes (SWCNT) by using neuronal PC12 cells [32]. The authors found that the toxicity of these two nanomaterials of relatively identical chemical structure was concentration-dependent, but, interestingly, these showed a different pattern of toxicity. FLG sheets resulted more toxic than SWCNT at low concentrations; whereas at higher concentrations graphene showed a lower activity, thus reversing its cytotoxic effect. These results can be primarily explained by the different shape of graphene sheets and SWCNT. Indeed, apart from concentration, the shape of nanomaterials plays an extremely important role, determining how these interact with cells and with potentially other biological systems.

Shortly after, another work evaluated the toxicity and biocompatibility of GO to human lung A549 cells, a widely used cell line for toxicity studies [33]. Compared to the previous study, the authors found that GO has much lower toxicity, as indicated by the results of the viability assay and Lactate Dehydrogenase (LDH) leakage assay. Together with MTT, the LDH assay, which measures enzyme release resulting from membrane damage, is another reliable and widely accepted test for evaluating materials toxicity [35]. The results of this work can be explained assuming that the cytotoxicity of nanomaterials is also heavily influenced by their surface functionalization. In this case, GO functionalization is higher than that of FLG sheets due to the presence of oxygenated functional groups. GO is indeed the oxidized derivative of graphene, produced by the oxidation of bulk graphite powders via chemical oxidation processes. It contains several reactive oxygen functional groups, such as epoxide, carboxyl and hydroxyl groups, which stabilizes the GO sheets on water, thus making them more hydrophilic [36].

Another work that stresses this concept is the one by Sasidharan and colleagues [37]. They found that, while pristine graphene accumulated on the cell membrane causing high oxidative stress leading to apoptosis, hydrophilic functionalized graphene, although internalized by the cells, did not cause any toxicity. The potential cytotoxicity of graphene-based nanomaterials is, therefore, highly dependent on their functionalization degree, as also established in the study of Das and coworkers [38]. When comparing the toxicity of GO and rGO, the first was found to be more toxic than rGO of same size on all the three cell lines used in the study. rGO is the reduced form of graphene; it is mainly produced to restore the electrical conductivity and optical absorbance of GO while reducing the oxygen content, surface charge and hydrophilicity [39]. Authors supposed that cytotoxicity was mediated by inducing ROS generation in the cells and that the presence of more reactive functional groups of GO would have a greater potential to interact with biological macromolecules compared with rGO. Nevertheless, by decreasing extent of oxygen functional group density on the GO surface, it was possible to reduce oxidative stress and consequently the nanomaterials toxicity.

The physicochemical mechanism of the oxidation-state dependent cytotoxicity of GO was proposed recently by the group of Zhang [40]. They synthesized three GO samples with similar size distribution and solubility in water, but with different oxidation state; these samples were then tested on Mouse Embryo Fibroblasts (MEFs). First of all, their results suggested that 50 μg/mL may be a threshold for GO to exhibit toxicity on normal mammalian cells. Then, they found that the GO with lower degree of oxidation displayed stronger toxicity on MEFs and stimulated higher intracellular ROS production (Figure 2a). As resulted from Electron Spin Resonance (ESR) spectrometry analysis, the decrease in the oxidation state was correlated to a higher ability of the GO samples to promote H_2O_2 decomposition and ·OH formation. Among various types of ROS, ·OH is an extremely reactive free radical and indiscriminately induces oxidative damages on various biomolecules, including DNA, lipids and proteins [41].

Soon after, another distinct process contributing to the molecular basis of graphene cytotoxicity was unraveled by Duan and colleagues [42]. They demonstrated that GO nanosheets were responsible to induce pore formation on cell membranes. Molecular dynamics simulations revealed that the molecular mechanism for perforation was dependent on cooperative lipid extraction driven by several graphene nanosheets (Figure 2b).

With regard to in vivo experiments, there are few studies concerning the biocompatibility of graphene-based nanomaterials since most of the investigations have only focused on the in vitro aspect of research. Nevertheless, the in vitro cell culture environment is definitely different from the in vivo complex 3D physiological condition. In addition, typical cytotoxicity assays measure the effect only during the first 12–24 h after exposure to a toxic substance although many biological reactions in vivo continue beyond 24 h.

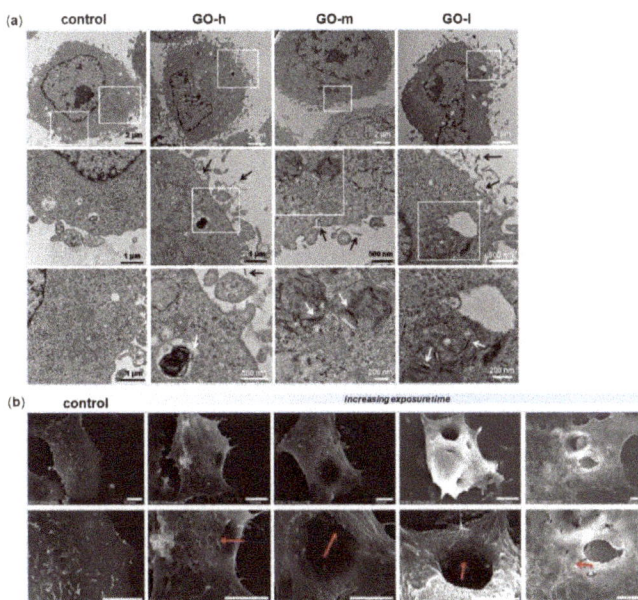

Figure 2. Morphological observations of cells after interactions with graphene-based nanomaterials. (**a**) Transmission Electron Microscopy (TEM) images of Mouse Embryo Fibroblasts (MEFs) treated with GO-high (GO-h), GO-medium (GO-m) and GO-low (GO-l) at 50 µg/mL for 24 h. On bottom, high-magnification images of the boxed-in photos on top are represented. The white and black arrows indicate GO aggregates inside and outside cells, respectively. (**b**) Scanning Electron Microscopy (SEM) images of cell membrane damage incurred by A549 cells as a result of GO nanosheets exposure observed during different phases of incubation. On bottom, high-magnification images of the boxed-in photos on top are represented. Reproduced with permissions from [40,42].

As for in vitro experiments, the in vivo effects of graphene-based nanomaterials were observed to be dependent on their concentration, physicochemical properties, exposure time, but also on administration route and on the characteristics of the animals used in the assays [43].

Wang and colleagues observed that intravenous injection of GO at 0.1 mg and 0.25 mg showed no obvious toxicity to mice, whereas almost half animal population died after administration of 0.4 mg GO [44]. All deaths occurred within days after injection of the GO. When investigating the effects of GO on organs of mice, the authors determined that graphene-based nanomaterials mainly accumulate in liver, spleen and lung after intravenous injection.

In another study, GO or pluronic-dispersed graphene was administered by intratracheal instillation to mice [45]. The authors found that GO induced acute lung injury that persisted for more than 21 days after administration. Lung inflammation was attributed to GO ability to increase the rate of mitochondrial respiration and the generation of ROS in cells, activating inflammatory and apoptotic pathways. On the contrary, toxicity was significantly reduced when the unoxidized graphene was well-dispersed with the block copolymer pluronic. These results demonstrate that graphene oxidation contributes to its pulmonary toxicity, whereas dispersion of pristine graphene in pluronic represents a safe way for handling and potential biomedical application of graphene.

Later, the group of Yang tested several Polyethylene Glycol (PEG) functionalized GO materials with varied sizes and surface chemistry after oral and intraperitoneal injection to mice [46]. The authors observed that few h after being administered by oral route, PEGylated GO derivatives were present in high quantities in stomach and intestine, but not in other major organs. However, only low levels were detected in the organs after 1 day and no trace was detected after 1 week. In contrast, high accumulation of PEGylated GO derivatives was observed in the liver and spleen after intraperitoneal injection, but their levels gradually decreased.

Considering certain inconsistencies and contradictory results in several studies, generalizations about graphene toxicity should be avoided. Indeed, it is generally accepted that the in vitro and in vivo behaviors of graphene-based nanomaterials are closely associated with their physicochemical properties, such as surface functionalization, coating, size and importantly, the administration routes. However, based on the studies presented in this section, we may summarize that the hydrophobic forms of graphene-based nanomaterials, which accumulate on the surface of cell membranes, are more toxic compared to the most hydrophilic ones. These, in fact, are more prone to infiltrate the cell membrane and to be internalized, as well as being removed from the application site. At the same time, when comparing GO with its reduced counterpart, rGO, the latter results less toxic than the hydrophilic form. Anyway, by controlling the GO reduction and maintaining the solubility, it is possible to minimize the GO toxicity and unravel the wide range of biomedical applications.

2.2. Graphene, Stem Cells and Osteogenesis

As mentioned before, one of the key characteristics of the ideal scaffold is an effective and controlled guidance of the proliferation and differentiation of stem cells into specific tissue lineages, either in the absence or presence of chemical inducers and growth factors. In the past few years, graphene-based nanomaterials have been increasingly used as scaffold materials for stem cell growth and differentiation. In the specific field of dental tissue engineering, most of the studies focused on evaluating the role of graphene in driving the osteogenic differentiation of stem cells. Indeed, bone regeneration plays a fundamental role in regenerative dentistry and how incorporation of graphene-based nanomaterials can enhance osteoconductivity through stimulating both cellular osteogenic differentiation and biomineralization have been extensively investigated [24].

In one of the first in vitro studies using stem cells, graphene produced by the Chemical Vapor Deposition (CVD) method and combined with different substrates was found to increase human Mesenchymal Stem Cells (MSCs) proliferation with no signs of cytotoxicity [47]. MSCs are one of the major progenitor cells achievable from a number of sources including adipose tissue, bone marrow and dental tissues [48]. Remarkably, graphene-coated scaffolds not only supported the growth and proliferation of human MSCs, but also accelerated their specific differentiation into osteoblasts, based on the level of the osteogenic marker osteocalcin and calcium mineralization level. The observed phenomena were obtained in the absence of additional growth factors and the acceleration rate was similar to that achieved with the use of typical osteogenic growth factors.

Starting from these results, another group investigated whether the graphene-induced osteogenesis of human MSCs was correlated to the chemical properties of graphene [49]. In the study, the effects of graphene and GO substrates were explored on both the osteogenic and adipogenic differentiation of human Bone Marrow MSCs (BM-MSCs). The results of the work revealed the ability of graphene to induce BM-MSCs differentiation into osteoblasts, probably by acting as a pre-concentration platform for dexamethasone and beta-glicerophosphate, typical osteogenic differentiation factors. In contrast, adipogenic differentiation resulted suppressed on graphene but strongly enhanced on GO substrates. In this case, the authors hypothesized that GO possesses high adsorption capacity of insulin, the key mediator for fatty acid synthesis.

Other studies have further tested the effects of GO on MSCs growth and osteogenic differentiation and the results varied. In the interesting work of Wei and colleagues, the authors compared the cellular behaviors of BM-MSCs towards pristine GO nanosheets using two in vitro biomimetic culture methods, that imitate similar situations in vivo [50]. The sequential-seeding method would mimic the interaction between GO and established cells, whereas the co-seeding method resembles the interaction between GO and migrating cells. The authors found that both cell proliferation and differentiation were dependent on pristine GO nanosheets concentration and in vitro culture methods. In particular,

BM-MSCs cultured with the sequential-seeding method resulted less vulnerable than those cultured with the co-seeding method, in which the cells had not yet adhered to the substrate. This seeding approach was beneficial for cells adhesion and proliferation when treated with 0.1 μg/mL of GO. Analogously, differentiation into osteoblasts was efficiently promoted when BM-MSCs were treated with this GO concentration, but only when osteogenic inducers were added to culture media (Figure 3a).

In recent years, also 3D graphene-based scaffolds have been actively explored and developed for the growth and differentiation of stem cells. The group of Crowder was the first to employ 3D graphene foams as substrates for human BM-MSCs culture [51]. When seeded onto these foams for 7 days, BM-MSCs were shown to maintain viability; in addition, the cells strongly expressed the osteogenic markers Osteocalcin (OCN) and Osteopontin (OPN), indicating their spontaneous osteogenic differentiation without the need for extrinsic biochemical factors (Figure 3b). It could be assumed that 3D graphene foams provided in vivo-like microenvironments which were conducive for the differentiation of stem cells.

Besides being used in their pristine form, graphene nanomaterials have also been combined with different polymers, nanofibers, or nanoparticles to create novel composites for controlling the growth and differentiation of stem cells. Hydroxyapatite (HAp), for example, is a calcium phosphate ceramic commonly used for bone repair or regeneration due to its chemical similarity to that of natural apatite in bones [52]. The recent work of Lee and colleagues demonstrated that rGO-coated HAp nanocomposites directed spontaneous differentiation of the murine preosteoblastic MC3T3-E1 cells towards bone lineage even in the absence of osteogenic differentiation factors [53]. Nevertheless, the osteogenic activity mediated by the rGO-coated HAp composites was further enhanced when these agents were added to the culture medium (Figure 3c). The authors concluded that, as the rGO-coated HAp composites showed to be potent inducers of spontaneous osteogenic differentiation of human MSCs, their application in the orthopedic and dental field may be determinant.

Natural bone is a composite material in which inorganic HAp nanocrystals are orderly embedded in an organic matrix made of collagen [54]. Collagen sponges are clinically approved scaffolds for bone regeneration [55]. These provide soft microenvironment to MSCs; however, stem cells preferentially differentiate to osteogenic lineage when cultured on mechanically stiff substrates. In order to overcome this problem, the group of Kang modified collagen sponge scaffolds by covalently incorporating GO flakes [56]. The covalent conjugation of GO flakes to collagen scaffolds increased the scaffold stiffness, which was comparable to that of pre-mineralized bone matrix and did not cause cytotoxicity of human MSCs. Importantly, when compared to non-modified collagen scaffolds, the GO-collagen sponges significantly enhanced the osteogenic differentiation of the cells. The authors hypothesized that the enhanced osteogenic differentiation was likely mediated by MSCs mechanosensing, since molecules that are involved in cell adhesion to stiff substrates, such as focal adhesions and cytoskeletal proteins, were either up-regulated or activated. In a following work of Nishida and coworkers [22], the effect of GO incorporation into 3D collagen sponges was examined in vitro and in vivo and compared to that observed with untreated collagen scaffolds. A previous study of the same group revealed clear morphological changes of collagen scaffolds after GO coating, that were dependent on GO concentration [57]. Notably, GO coating did not alter the porous structure of collagen sponges but reinforced their stability. Also, in their following work, the authors found that GO application improved the physical properties of collagen scaffold, such as compressive strength, enzyme resistance and adsorption of calcium and proteins. In addition, when GO scaffolds were implanted into a dog tooth extraction socket, the rate of new bone formation was higher than in the control group consisting of untreated collagen scaffolds, where most of the socket area was filled by connective tissue. These results were obtained by means of histomorphometric measurements and radiographic images taken 2 weeks after implantation.

Among others available materials, peptides containing the repeating arginine-glycine-aspartic acid (RGD) sequence, which is often found in Extracellular Matrix (ECM) proteins, have been widely applied to enhance cell adhesion to synthetic polymers or metallic surfaces [58,59]. In one of the latest studies, a 3D graphene/RGD peptide nanoisland composite was developed and tested on the osteogenesis of human Adipose-Derived Stem Cells (ADSCs) [60]. Human ADSCs represent excellent candidates being accessible for autologous transplantation and able to differentiate into osteogenic lineage for bone regeneration [61]. Also, in this work beta-glicerophosphate, dexamethasone and ascorbic acid were added to the culture medium for induction of osteogenesis. Such scaffolds were found to accelerate ADSCs osteoblastic differentiation, as revealed by gene expression results, Alkaline Phosphatase (ALP) activity measurements and evaluation of calcium deposition over a period of 4 weeks. Thus, the combination of RGD-containing peptides and GO substrates enhanced the adhesion strength of stem cells, ultimately resulting in increased osteogenic differentiation, as cell adhesion is the first step driving osteogenic lineage commitment. Taken together, the results of these studies would suggest that nanomodifications of different scaffolds using GO might provide good structure for colonization by several cell types.

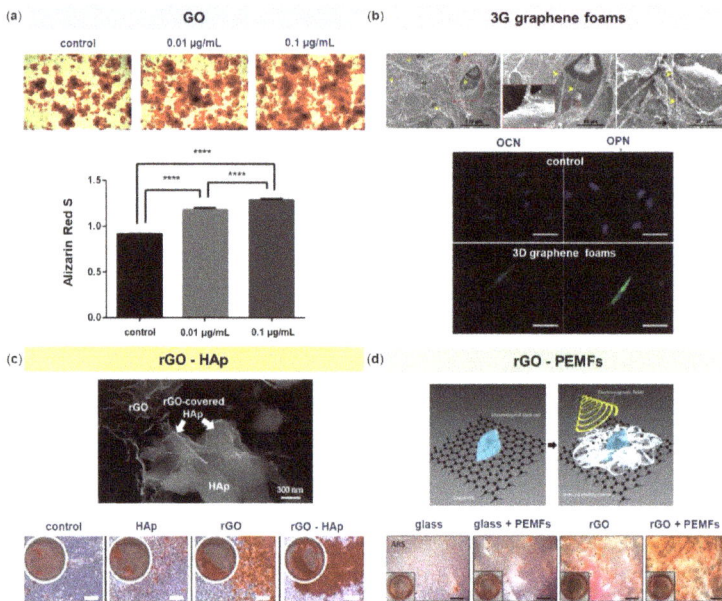

Figure 3. Effect of graphene-based nanomaterials on the osteogenic differentiation of Mesenchymal Stem Cells (MSCs). (**a**) Evaluation of matrix mineralization by means of Alizarin Red S (ARS) staining (top panel) and quantification (bottom panel). MSCs were grown for 21 days in osteogenic differentiation medium on GO nanosheets. The level of mineralization in the 0.1 μg/mL GO group was significantly higher than the other two conditions. (**b**) Human BM-MSCs cultured on 3D graphene foams for 4 days show protrusions up to 100 μm in length (yellow arrowheads) that extended from the cell bodies (black arrows), as evidenced by SEM images (top panel). These 3D substrates were also found to promote the expression of the osteogenic markers Osteocalcin (OCN) and Osteopontin (OPN), as displayed by immunofluorescence images (bottom panel). (**c**) SEM images of the rGO-coated Hap nanocomposites showing that Hydroxyapatite (HAp) particles were partly covered and interconnected by an network of rGO nanosheets (top panel). ARS staining performed at 21 days reveals that rGO-coated HAp nanocomposites significantly increase calcium deposits in MC3T3-E1 cells compared to the non-treated control and rGO or HAp alone (bottom panel). (**d**) Schematic representation of the study: human BM-MSCs were seeded onto rGO substrates, then exposed to Pulsed Electromagnetic Fields (PEMFs) (top panel). The rGO+PEMFs group exhibited the strongest staining as evidenced by ARS staining performed after 2 weeks from cells seeding. Reproduced with permissions from [50,51,53,62].

One last work worth mentioning is the one of Lim and colleagues [62]. The authors used Pulsed Electromagnetic Fields (PEMFs) in combination with rGO substrates to evaluate the osteogenic differentiation of human alveolar BM-MSCs. PEMFs therapy is based on the physical effect sorted from the combination of an electric with a magnetic field, generating waves that propagate at the speed of light; and it has already been reported to strongly enhance the commitment of ADSCs into mature osteoblasts in vitro [63]. Cells cultured using osteogenic medium were exposed to 30 minutes daily treatments of 0.6 ± 0.05 mT and 50 Hz PEMFs for up to 2 weeks. After 7 days, rGO+PEMFs promoted the expression of ALP, an early marker of osteogenesis, whereas at week 2, the same scaffolds stimulated OPN, OCN and Runt-related transcription factor 2 (RUNX2) expressions as well as calcium deposition. In addition, DNA microarray analysis established that the combination of rGO and PEMFs increased the expression of genes related to ECM formation, membrane proteins and metabolism. The results of this work seem to demonstrate the synergic effects of rGO scaffolds and PEMFs on the osteogenesis of human alveolar BM-MSCs (Figure 3d).

All the works presented in this section demonstrate that, over the last few years, graphene has been found as one of the most promising biocompatible scaffolds for human MSCs adhesion, proliferation and differentiation, particularly towards the osteogenic lineage. These recent investigations clearly indicate a potential or active ability of graphene and its derivatives to induce the osteogenic differentiation of MSCs. Graphene alone seems to drive the spontaneous osteogenesis of stem cells; however, the presence of osteogenic inducers or growth factors in culture medium contributes in promoting this effect. Further summarizing the studies presented here, we may conclude that the association of graphene with stiff substrates, such as HAp, or with natural bone components, such as collagen, enhances the intrinsic properties of graphene in guiding the osteogenic differentiation of stem cells. Also mimicking the natural bone ECM, for example through the production of 3D graphene-based scaffolds, has proved to be a novel strategy in directing the stem cell fate towards the osteogenic lineage. Nevertheless, a deeper understanding of the ability of graphene to improve the biological properties of different scaffold materials will be essential to engineer graphene-based substrates for targeted biomedical applications.

Regarding stem cells of dental origin, few data are available to date. Dental tissues are a very rich source of stem cells, which are collectively called Dental Stem Cells (DSCs) [9]. In the following paragraphs, studies describing applications of graphene-based nanomaterials in association with DSCs will be presented.

2.2.1. Dental Pulp Stem Cells (DPSCs)

Dental pulp is the soft connective tissue of the tooth and Dental Pulp Stem Cells (DPSCs) are a unique MSCs population that is present in the cell rich zone and core of the pulp. These cells were firstly discovered in 2000 by Gronthos and colleagues, thanks to their ability to differentiate into odontoblasts/osteoblasts, adipocytes and neural-like cells [64].

Previous studies established that DPSCs and BM-MSCs are analogous cell populations, since they share expression profile for thousands of genes, among which those associated with the initiation of mineralization and bone homeostasis [65]. Other commonly expressed genes include those coding for various growth factors known to be strong promoters of osteogenesis, such as Fibroblast Growth Factor 2 (FGF-2) and Bone Morphogenetic Protein 2 (BMP-2) and BMP-4, but also genes encoding ECM components such as ALP, Collagen type I (COLI) and OCN. Human dental pulp is an optimal source of DPSCs due to its easy surgical access, the very low morbidity of the anatomical site after the collection of the pulp and the high efficiency of the extraction procedure of the stem cells from the tissue [66]. In addition, compared to equal volumes of bone marrow, dental pulp contains a higher number of MSCs.

In the study of Rosa and coworkers, the effect of GO-based scaffolds on DPSCs proliferation and differentiation was evaluated and compared to that of glass substrates [67]. Cell morphology and proliferation were evaluated by SEM and fluorescence microscopy after 5 days from seeding; mRNA expression of Msh homeobox 1 (MSX-1), Paired box 9 (PAX-9), RUNX2, COLI, Dentin Matrix acidic Phosphoprotein 1 (DMP-1) and Dentin Sialophosphoprotein (DSPP) genes was measured at 7 and 14 days. The results of this study indicated that DPSCs were able to adhere on both the glass and GO surfaces, without any significant difference in proliferation. Nevertheless, when mRNA expression levels were measured, these resulted significantly higher for all genes tested onto GO compared to glass surfaces after 14 days from seeding (Figure 4a). MSX-1 is essential for the development of teeth, craniofacial structures and/or limb structures in embryos; it is a target activator of PAX-9 and the deletion of them causes tooth agenesis in mice [68,69]. RUNX2 is a transcription factor that is essential for osteogenic differentiation and can up-regulate the expression of OCN [70]. DMP-1 is highly expressed during odontogenesis and in mature odontoblasts, it is capable to produce dentin in vivo and its deficiency leads to dentinogenesis imperfecta [71]. DSPP codes for two proteins, dentin sialoprotein and dentin phosphoprotein, both of which have been associated with early dentinogenesis and are absent in bone [72]. What is really considered interesting in this work is the fact that the GO substrates provided an up-regulation for all the measured gene expression, even if no inducers for differentiation were used. In addition, the substrate was also capable of increasing the expression of both DMP-1 and DSPP, which are intimately related to odontogenic differentiation of stem cells from dental pulp. This means that the use of GO substrates by themselves may enhance the expression of odontogenic genes opening new opportunities to the use of GO alone or in combination with dental materials to improve their bioactivity and beyond.

In another study, the authors aimed at evaluating the potential of graphene to induce odontoblastic or osteogenic differentiation of DPSCs without the use of any chemical inducers [73]. Dentin and bone are considerably made of a HAp that is associated with the matrix produced by odontoblasts and osteoblasts, respectively [64]. Mineralization, odontogenic and osteogenic genes expression and protein expression of RUNX2, COLI and OCN were evaluated in DPSCs seeded on graphene and glass substrates for 14 and 28 days. Regarding the mineralization process, the graphene substrate positively influenced the cells to spontaneously secrete mineralized matrix. When comparing the gene expression levels, the odontoblastic-related genes MSX-1, PAX-6 and DMP-1 were significantly down-regulated on graphene substrates compared to glass. On the contrary, graphene increased both gene and protein expressions of RUNX2 and OCN, thus suggesting its ability to induce osteogenic differentiation rather than odontogenic differentiation of DPSCs. An explanation to this may be the high Young modulus of graphene (1.0–2.4 TPa) [74], which can contribute to the osteogenic differentiation of DPSCs as these cells usually need soft substrates to achieve the odontogenic differentiation [75]. Taken together, the results of this study suggest that DPSCs are not prone to differentiate into odontoblastic-like cells when grown on graphene, but this substrate is able to induce osteogenic differentiation of the cells without the use of chemical inducers for osteogenesis.

Therefore, authors seem to agree with the fact that graphene is a biocompatible material for DPSCs and it might be used as a valid substrate for dental tissue engineering since it is able to promote either osteogenic or odontogenic differentiation of DPSCs depending on the surface characteristics and functionalization of graphene, as also emerged in the previous section.

2.2.2. Periodontal Ligament Stem Cells (PDLSCs)

The periodontium refers to those structures which surround and support teeth. It is composed by four parts: gingiva, alveolar bone, cementum and periodontal ligament. The function of the periodontal ligament is to literally attach the teeth to the bone and when a chronic process, called periodontitis, affects these structures, the teeth stability and health are compromised. All the periodontium is constantly maintained by Periodontal Ligament Stem Cells (PDLSCs), which have the ability to differentiate into cementoblasts, odontoblasts and fibroblasts [76].

Thanks to their potential to differentiate into several types of specialized cells, PDLSCs can be used as substitute to BM-MSCs in order to determine the osteogenic potential and the viability onto different substrates [77]. Even if in literature the bone formation potential of PDLSCs on various surfaces such as Titanium (Ti) has been already demonstrated [78], few articles evaluated the biocompatibility of graphene-based nanomaterials for human PDLSCs.

In one of the few studies, the effects of GO, Silk Fibroin (SF) and GO combined with SF were investigated on PDLSCs adhesion, proliferation, viability and expression of MSCs markers [79]. SF is widely used as a biocompatible material in the fabrication of cellular scaffolds for tissue engineering and a composite of the two materials has already been proposed for several biomedical applications [80,81]. Nevertheless, its performance as a substrate for growing of PDLSCs has not been addressed yet. In this work, PDLSCs obtained from healthy extracted molars were used and cultured up to 10 days on the above-mentioned substrates and on plastic, which represented the control condition. Cell adhesion was higher on GO and on the GO-SF composite film rather than on fibroin alone, as revealed by immunofluorescence staining of the actin cytoskeleton (Figure 4b). Regarding the cells proliferation rate, the results of the MTT assay showed a high rate of proliferation for PDLSCs growing on GO films compared to plastic and on GO and fibroin composite film compared to fibroin alone after 10 days from seeding. In order to evaluate if the biomaterials employed in this study were able to maintain the mesenchymal phenotype of PDLSCs, flow cytometry was used to assess the expression of some MSCs surface molecules. Culture of PDLSCs on SF, GO or GO-SF composite did not significantly alter the level of expression of the typical MSCs markers CD73, CD90 or CD105 compared to expression levels displayed by PDLSCs cultured on plastic The authors concluded that the GO-SF composite significantly improved the performance of the fibroin film, as an alternative to the coating with collagen, representing an interesting combination of biocompatibility, induction of proliferation and mechanical resistance, very suitable for working in cellular environments where mechanical resistance is required.

Later, the same group investigated the potential of SF and GO composites to promote human PDLSCs spontaneous differentiation into osteo/cementoblast-like cells [82]. In the study, the authors have optimized the parameters of fabrication of the GO-SF composite film, using different ratios of the two materials and also varying the graphene oxidation status, with the aim to find the best configuration for cell proliferation and differentiation. The PDLSCs proliferation rate was consistently improved in those combinations containing low amounts of graphene and a high SF dose, as emerged from the MTT assay results at 7 and 10 days of culture. The authors concluded that the best configurations in terms of PDLSCs proliferation were GO alone and rGO:rSF at 1:3 ratio. SF was used to confer 3D characteristics to the GO or rGO as well as to improve its handling. Previous data have shown that human PDLSCs bioengineered on 3D graphene scaffold preparations are associated with higher proliferation rates than on 2D ones [83]. Flow cytometry analysis was further used to confirm the mesenchymal phenotype of the isolated PDLSCs and to determine possible phenotypic changes after their culture on the different graphene-fibroin combinations. The MSCs surface molecules CD73, CD90 and CD105 were found to be present in the PDLSCs grown on all the biomaterials tested, although their expression decreased with culturing time. This is not surprisingly, since the expression level of MSCs markers progressively declines in stem cells during their multilineage differentiation process [84,85]. After demonstrating the beneficial effects of GO-SF composites on PDLSCs proliferation, gene expression analyses were performed to further characterize the effect of these scaffolds on PDLSCs differentiation into osteo/cementoblast-like cells. No osteogenic chemical inducers were added to the culture medium. GO-SF composites, particularly the reduced configurations rGO, rSF and rGO-rSF, were found to induce the over-expression of early osteoblast/cementoblast markers, including BMP2, RUNX2, ALP and COLI. On the contrary, Osterix (OSX) and Osteocalcin (OCN), whose expressions are associated with late osteoblast differentiation stages, were found to be down-regulated on all the substrate preparations. Starting from this result, the authors explored the expression of two specific cementum-related genes, Cementum Attachment Protein (CAP) and Cementum Protein 1 (CEMP1), that are expressed at early and late differentiation stages, respectively [86]. All the scaffolds tested were associated with significant down-regulation of CAP and concomitant over-expression of CEMP1, suggesting that graphene-fibroin composites can induce cementoblast differentiation of the human PDLSCs in the absence of any growth factors.

2.2.3. Dental Follicle Progenitor Cells (DFPCs)

The dental follicle is the loose connective tissue surrounding the enamel organ and the dental papilla of the developing tooth germ. Its principal role is the coordination of tooth eruption through the regulation of the osteoclastogenesis and osteogenesis processes [87]. The Dental Follicle Progenitor Cells (DFPCs) are multipotent stem cells that have immunomodulatory properties, high proliferation rate and ability to differentiate into odontoblasts, cementoblasts, osteoblasts and other cells implicated in the teeth [88]. Furthermore, they are able to re-create a new periodontal ligament after in vivo implantation.

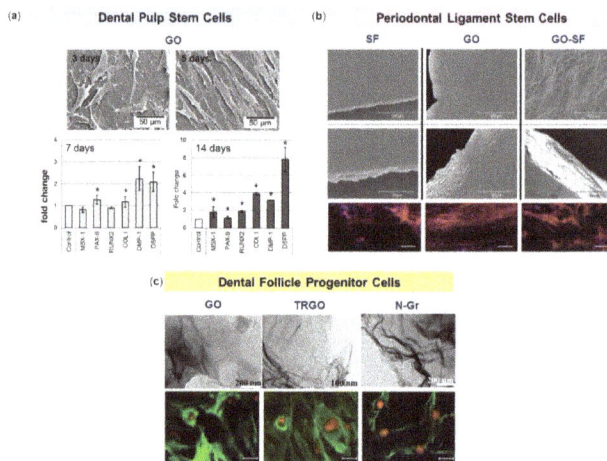

Figure 4. Interactions of graphene-based nanomaterials with Dental Stem Cells (DSCs). (**a**) SEM images showing that Dental Pulp Stem Cells (DPSCs) can efficiently adhere and proliferate on GO substrates for 3 and 5 days (top panel). DPSCs on GO present higher expression compared to glass (control) for all genes tested both at 7 and 14 days (bottom panel). (**b**) SEM images of films composed of Silk Fibroin (SF), GO and GO-SF mixture (3:1) at different magnifications (top panel). Immunofluorescence staining of the actin cytoskeleton showing a higher adhesion of Periodontal Ligament Stem Cells (PDLSCs) on GO and on the GO-SF composite film rather than on SF alone at 7 days. (**c**) TEM images of GO, Thermally Reduced Graphene Oxide (TRGO) and Nitrogen-doped graphene (N-Gr) (top panel). Confocal microscopy images of human Dental Follicle Progenitor Cells (DFPCs) seeded on GO, TRGO and N-Gr at 40 µg/mL showing staining of cytoskeleton actin filaments (green) and nuclei (red) (bottom panel). Reproduced with permissions from [67,79,89].

To date, only one study evaluating the behavior of human DFSCs on graphene-based nanomaterials has been reported [89]. In the work, human DFSCs obtained from healthy extracted teeth were seeded onto GO, Thermally Reduced Graphene Oxide (TRGO) and Nitrogen-doped graphene (N-Gr) substrates. Then, cytotoxicity, oxidative stress induction, cellular and mitochondrial membrane alterations were analyzed for all the developed substrates (Figure 4c). The results of this study showed that GO was the less toxic for the cells, followed by N-Gr, whereas the TRGO substrate resulted the more toxic. Regarding the oxidative stress level, both the GO and the TRGO substrates induced lipid peroxidation without significant alteration of the membrane. On the other hand, the N-Gr substrate, which showed to have a positive antioxidant effect on the stem cells studied, reduced cell viability without oxidative stress. The authors suggested that N-Gr affected the cells not only determining a physical damage of the plasma membrane due to graphene irregularities, but also by the ability of nitrogen to coordinate biomolecules thus interfering in the biological processes. As regards the mitochondria membrane potential, it was decreased in all the substrates used in this study following a dose-dependent manner. This effect was higher in the TRGO than in the GO substrate, while in the N-Gr substrate structural lesions were present only at high concentrations (20 and 40 µg/mL). The authors concluded that while GO and N-Gr may be considered as promising fillers for various dental nanocomposites, TRGO does not seem to be a suitable material.

2.3. Graphene and Immune Cells

A critical step for the application of nanotechnology in medicine is the interaction of a material with the immune system because, once administrated, it immediately enters in contact with the immune cells. Graphene and its composites are not exempt from the interaction with cells of the immune system and, as in the last few years the biological applications of graphene-based nanomaterials have grown, it results necessary to investigate the immune-related impact of these nanomaterials. Many papers about this issue underlined stimulation or suppression effects of graphene-based nanomaterials on immune system cells, depending mainly to graphene purity, oxidation level, shape dimension, functionalization and to the type of immune cells analyzed.

The first extended investigation on graphene and immune cells was made by Sasidharan and coworkers [37]. After comparing the effects of pristine graphene and GO on Peripheral Blood Mononuclear Cells (PBMCs), the authors observed really different effects: pristine graphene induced a significant release of Interleukin-8 (IL-8) and IL-6 whereas GO determined only a moderate release of IL-8, despite both seemed to have an excellent compatibility with PBMCs.

To best characterize the impact of GO on immune cells, Orecchioni and colleagues studied and compared the effects of small-size (100–500 nm) and large-size (1–10 µm) GO flakes on a pool of PBMCs [90]. The authors detected a more efficient cellular activation and cytokine release by small-size GO. In particular, through gene expression analysis, they underlined that small GO is able to modulate a large number of genes related to cytokines and activation-related genes, such as T-box transcription factor (TBX21) and CD80. Furthermore, it was found that GO promotes the over-expression of many pathways such as leukocytes chemotaxis pathway and genes associated with C-X-C motif chemokine 10 (CXCL10) ligand and C-X-C motif chemokine receptor 3 (CXCR3), able to activate T-helper and Natural Killer (NK) cells. Once established various effects on the total PBMCs, several studies have then been performed also analyzing the effect of graphene on specific immune cell types, as described in the following paragraphs.

2.3.1. Lymphocytes

Lymphocytes, which represent about the 20–40% of the white blood cells and the 70–80% of PBMCs, are responsible for the antigen-specific and innate immune response. The most of research have been made analyzing the effects of graphene-based nanomaterials on T lymphocytes, concluding that the cytotoxicity of graphene depends on its dosage, time of exposure and type of material with which these are combined. For example, Ding and coworkers demonstrated that low doses of GO have no cytotoxic effects on T lymphocytes, while a dose over 100 µg/mL causes oxidative stress related apoptosis after 24 h of incubation [91]. In particular, at high concentration GO was found absorbed on cell membrane, without internalization or cell membrane disruption, but it caused a reduction of cell viability and an increase of lymphocytes apoptosis probably due to the production of ROS. However, no significant reduction in T lymphocytes immune response was found. The same tests conducted on GO-polyethylenimine (GO-PEI) revealed that this kind of functionalization severely damaged T lymphocytes and suppressed their immune ability [91]. On the contrary, the functionalization of GO with polyvinylpyrrolidone (PVP) increased the biocompatibility of GO and at high doses of GO-PVP the amount of apoptosis was much less then in GO alone [92].

2.3.2. Macrophages

Half of the studies about graphene and the immune system have been carried out on macrophages, not only because these represent the first phase of body immune reaction, but also because macrophages are the easiest long-term culturing primary immune cell population [93].

The phagocytic activity of macrophages makes them the most exposed cells to the negative influence of graphene and its derivatives [94]. It has been reported that the ability of macrophages to the internalization mainly depends on the size of the sheets: the smaller is the lateral dimension of GO, the greater the ability of macrophages to internalize it [95]. Moreover, the capacity of macrophages to phagocyte GO can be modulated by GO functionalization. Zhi and colleagues demonstrated that the functionalization of GO with PVP decreases the internalization of the compound, indicating that PVP-GO has a better immunological biocompatibility (Figure 5a) [92].

Regarding the activation of macrophages, Chen and colleagues showed that pristine graphene and GO promoted the production of several cytokines, such as IL-1α, IL-6, IL-10 and Tumor Necrosis Factor alpha (TNF-α), via NF-κB pathway, that is activated via Toll-Like Receptors (TLRs) [96]. Graphene promotes the activation of IκB Kinase (IKK), thereby activating NF-κB [97,98]. Moreover, GO stimulates the production of Myeloid Differentiation primary response gene 88 (MyD88), another protein related to TLRs. Activated TLRs start the kinase cascade, by a MyD88 mechanism, that promotes the translocation of NF-κB into the nucleus, thus increasing cytokines gene expression (Figure 5b). In addition, an elevated activation of TLRs, especially TLR-4, causes TLR-dependent necrosis of macrophages. The silencing of genes that code for TLRs, completely immunize macrophages from graphene [99]. The stimulation of TLRs also favors the formation of phagosomes that active enhance the uptake of graphene and the consequent affection of cell metabolism and gene/protein expression [99]. Experiments on phagocytic process showed that 100 μg/mL GO generate the same effects of LPS stimulation that is the activation of macrophages and the production of pro inflammatory cytokine [100].

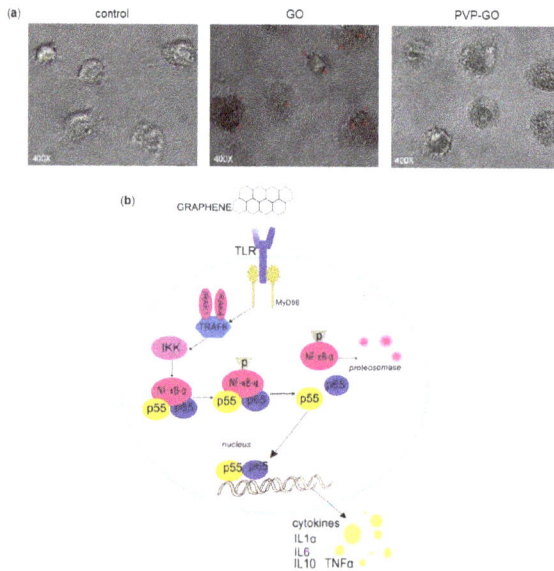

Figure 5. Interactions of graphene with macrophages. (**a**) Optical micrographs of macrophages treated with 25 μg/mL GO and PVP-GO for 48 h. Macrophages showed to be inclined to GO internalization (red arrows), while the functionalization with PVP prevents the phenomenon. (**b**) Signaling pathway of macrophage activation stimulated by graphene. Graphene may stimulate Toll-Like Receptors (TLRs), thus activating kinase cascade Myeloid Differentiation primary response gene 88 (MyD88)-dependent mechanism. IKK activation initiates the phosphorylation and degradation of IκB and consequently, the release of NF-κB subunits and their translocation into the nucleus. NF-κB binds to the promoter regions of its effector genes and initiates the transcription of multiple pro-inflammatory genes and the secretion of Interleukin 1α (IL-1α), IL-6, IL-10, Tumor Necrosis Factor alpha (TNF-α). Reproduced with permissions from [92,99].

2.3.3. Dendritic Cells (DCs)

Dendritc cells (DCs), representing the 1–2% of PBMCs, play a fundamental role in the activation of antigen-specific T lymphocytes [101]. Few data are available about interactions of graphene-based nanomaterials and DCs. Tkach and coworkers showed that GO incorporated by DCs down-regulates the intracellular level of LMP7, a part of an immunoproteasome required for processing protein antigen [102]. In another study, Zhi and colleagues compared the effect of GO and PVP-GO on DCs, highlighting a capacity of GO to promote maturation of DCs and also the overexpression of IL-6 in a dose-dependent manner, compared to the non-treated group [92]. The authors concluded that functionalization with PVP mitigate the effects of GO, that resulting in a lower immunogenicity.

2.4. Graphene and Antibacterial Activity

Infection is a frequent process during biomaterial implantation procedures. Since antibiotics often impact negatively on bacterial flora and pathogens are able to acquire resistance against different antibiotics, in the last few years various nanotechnologies with antimicrobial properties have been developed and studied [103,104], trying to get closer to the ideal scaffold that should inhibit bacterial growth at the surface, while simultaneously promoting cell adhesion and proliferation [43].

It has been reported that some graphene-based nanomaterials possess antibacterial properties; nevertheless, the effects of graphene on bacteria structure, metabolism and viability depend on the materials' concentration, time of exposure, physical-chemical properties, as well as on the characteristics of bacteria used in the tests [105,106].

A lot of studies indicated that graphene and several graphene nanocomposites have a remarkable antibacterial ability against both Gram-negative and Gram-positive bacteria [107–110]. It has been demonstrated that the antibacterial effect is due to the capacity of graphene to physically damage microorganisms through different mechanisms: graphene nanostructure acts as a nano-knife penetrating and cutting cell membrane, it wraps cells inducing mechanical stress, it is able to extract phospholipids from lipid membranes and it produces oxidative stress through ROS generation but also by charge transfer phenomena. Molecular dynamic simulations suggest that thin graphene nanosheets can insert into both bacterial membranes creating a break and, once inserted, the Van Der Walls forces and the hydrophobic properties of graphene promote phospholipids extraction from the lipid layers of the bacterial membranes causing irreversible damages [111].

Another way for graphene to exert antibacterial activity seems to be through the separation of microorganisms from the microenvironment. Aggregated graphene sheets in suspension can isolate bacteria from the surrounding environment hindering nutrient consumption, reducing their ability to proliferate and favoring their inactivation [108].

Graphene antibacterial activity is also probably related to oxidative stress induction. This is mainly mediated by ROS generation, especially when GO is used, nevertheless oxidative stress can be triggered without ROS generation. Li and coworkers showed that graphene is able to act as an electron pump pulling out electrons from bacterial membranes and inducing ROS-independent oxidative stress that affects microorganisms [110].

Obviously, these mechanisms are not specific against bacteria but can also affect other cells, although with less effectiveness. Pang and colleagues tried to clarify the ratio between biosafety and antibacterial activity of graphene and GO [112]. They observed that both cytotoxicity and antibacterial effects are dose-dependent and that GO has a higher activity; in particular, a GO concentration in the range of 50–100 µg/mL keeps the balance between minor cytotoxic effects and major antibacterial activity.

Among graphene derivatives, GO attracted particular attention thanks to the ease with which it can be functionalized, representing the precursor of the most graphene nanocomposites, but also thanks to its excellent water dispersity. However, a lot of study established that antibacterial activity of GO is dependent not only on its concentration but also on the size and the lateral dimension of GO sheets. Liu and colleagues demonstrated that the antibacterial activity of GO nanosheets depends on their size [113]. Indeed, larger GO sheets express stronger antibacterial activity against *Escherichia coli* compared to smaller one, probably due to the capacity of larger GO sheets to completely cover bacteria inhibiting their proliferation and colony formation (Figure 6a).

In the environment of the oral cavity, *Streptococcus mutans*, *Porphyromonas gingivalis* and *Fusobacterium nucleatum* are the most representative bacteria responsible for caries, periodontal and periapical diseases [114]. The microbial community resident in the mouth exists in balance with the oral microenvironment [115]. Host susceptibility, diet and habits could lead to a break in balance that gives rise to adverse reactions. In particular, *S. mutans* is a Gram-positive facultative anaerobic bacterium importantly involved in caries formation and in the modification of the oral microenvironment, decreasing the pH value by the production of large amount of organic acids [116]. Instead, *P. gingivalis* and *F. nucleatum* are Gram-negative anaerobic bacteria, associated with periodontitis [117]. For these reasons, antibacterial activity of graphene and its nanocomposites, in particular against these cariogenic bacteria, has been studied. He and collaborators investigated the effect of GO against dental pathogen bacteria, showing that the viability of *S. mutans*, *P. gingivalis* and *F. nucleatum* decreased in the presence of GO nanosheets depending on its concentration in a dose-dependent manner [114]. TEM images clearly showed that, when the GO was present, the integrity of *S. mutans*, *P. gingivalis* and *F. nucleatum* was strongly compromised due to the severe insertion, cutting and destructive extraction of lipid molecules effect that GO act against the membrane (Figure 6b).

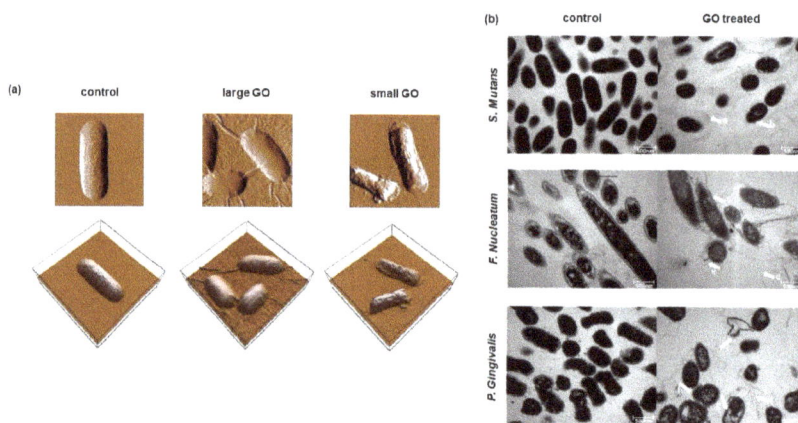

Figure 6. Effect of GO nanosheets on bacteria. (**a**) Atomic Force Microscopy (AFM) amplitude (top) and 3D (bottom) images of *Escherichia coli* cells 2 h of after incubation with/without GO sheets. *E. coli* cells incubated with deionized water without GO sheets show a preserved integrity of the membrane (control). The incubation with the 40 µg/mL large GO sheets suspension results in a completely cover of bacterium surface by GO sheets, whereas small GO sheets adhere to cell surface without fully covering it. Scale bars are 1 µm. (**b**) TEM images of *Streptococcus mutans*, *Fusobacterium nucleatum* and *Porphyromonas gingivalis* cells after incubation with GO nanosheets dispersion (right side) for 2 h and after incubation with saline solution for 2 h as control (left side). All treated cases had the same GO dose of 80 µg/mL. Scale bars are 500 nm. Reproduced with permissions from [113,114].

Graphene and some of its composites seem to exert their activity not only against single bacteria but also against bacterial biofilms. Biofilms are surfaced-attached bacterial communities that self-produce adhesive ECM; they play a role in a wide variety of infections, i.e., caries, catheter infection and bloodstream infection [118]. Several complicated and expensive methods to prevent biofilms formation have been proposed, including coating of nanomaterials with ion or polymers impregnated with antibiotics [119–122]. Recently, inspired by graphene antibacterial ability and its ease functionalization, possible effects of graphene-based nanomaterials against biofilms are being studied. For example, Song and coworkers investigated the influence of GO on bacterial biofilm formation, observing that high GO concentrations inhibit the formation of Gram-negative and Gram-positive biofilms via membrane stress, whereas low GO concentrations enhance their formation [123]. The authors hypothesized that low GO concentration kills only a limited part of bacteria and dead cells could serve as a protection barrier and nutrient to the rest of biofilm formation, whereas high GO concentration promotes the inactivation of most bacteria, hindering the biofilm growth. In another work, Mao and colleagues tested the antibiofilm activity of a GO-aptamer composite and compared to that of GO and aptamers per se [124]. They showed that all agents interacted with pathogen disturbing the initial growth of biofilm and destroying the established biofilm, but the combination of GO and aptamers exhibited a superior synergic effect than the single substances.

Despite a lot of studies stressed an antimicrobial activity for pristine graphene and GO, some other works evidenced that graphene has no intrinsic bactericidal properties, but even it is able to enhance bacteria growth when a colloidal suspension is formed [125,126]. Das and Ruiz in their studies showed that only the functionalization with antibacterial agents such as silver nanoparticles conferred to the graphene-based nanomaterials a bactericidal activity [125,126]. In agreement with these studies, Some and collaborators confirmed that graphene alone has no antibacterial properties, but graphene-iodide composites, above all double-oxidizes GO-iodide, have a potent antibacterial activity [127].

Wu and coworkers tried to clarify the debate concerning the antimicrobial activity of graphene and its derivatives comparing the antibacterial effects of GO, GO-polyoxyalkyleneamine (GO-POAA) and GO-chitosan against *E. coli* and *Bacillus subtilis*, Gram-negative and Gram-positive bacteria, respectively [128]. The results of this study indicated that less than 50 µg/mL GO in a nutrient medium solution has no antimicrobial activity; on the contrary, it enhances bacterial growth acting as a biofilm that allows bacterial attachment and proliferation. On the other hand, the conjugation of GO with POAA or chitosan, two antibacterial molecules, at the same concentration showed antibacterial effects. All materials demonstrated antibacterial activity when bacteria were grown in Phosphate Buffered Saline (PBS) solution.

All these studies would suggest that the antibacterial activity of graphene-based nanomaterials is variable and dependent on the conditions of the tests; in particular, results change depending on the type of material and the size of particles, on its concentration and its state (in solution or adsorbed on a support), on the type of bacteria used in the study and on the medium used for their growth.

3. Applications of Graphene-Based Nanomaterials in the Dental Field

Graphene and its derivatives display many potential applications in the dental field, as can be seen from the literature. Thanks to their potentiality, particularly in driving the osteogenic differentiation of stem cells and antibacterial abilities, the application of graphene-based nanomaterials to already existing dental technologies are being studied. In addition, graphene seems to be interesting as a platform able to release therapeutic molecules to improve implants osseointegration and bone formation. In the following paragraphs, we will present several studies discussing the association of graphene with dental implants, membranes, resins, cements and adhesives, as well as strategies for teeth-whitening.

3.1. Graphene and Dental Implants

Nowadays, Ti dental implants are considered as the best substitutes for missing teeth thanks to their reliability and predictability, besides to their mechanical strength and favorable biocompatibility [129,130]. Since these devices are in close contact with the surrounding tissues, a critical parameter for the success of the implantation is the ability of the implant to integrate with the tissue promoting the osseointegration process; this represents indeed a fundamental step for obtaining the best new bone quantity and quality [129,131]. During this process, a key role is played by the implant surface characteristics such as roughness, surface treatment and hydrophilicity.

Ti inertness represents a disadvantage because it may induce the development of fibrous tissue which could lead to the failure of the implant [130]. For this reason, several studies have focused on modifying implant surfaces to favor a better osseointegration. In this context, graphene appeared to be an excellent implant-coating for hard tissue engineering in order to accelerate bone regeneration.

As evidenced earlier, GO-coating was found to be a successful substrate for anchoring and growing of DPSCs on its rough surface and for promoting their osteogenic differentiation through up-regulation of MSX-1, PAX-9, RUNX2, COLI, DMP-1 and DPSS in particular after 14 days of culture [67]. To improve stem cells osteogenic differentiation, the groups of Jung and Ren functionalized GO-Ti implants using different methods with dexamethasone, a synthetic glucocorticoid that is known to contribute to this phenomenon [130,132]. Independently of the method of functionalization and coating, both groups showed that GO-coating of Ti implants, even more when functionalized with dexamethasone, improves implants biocompatibility, cell proliferation and mostly cell osteogenic differentiation.

Another approach to improve the osseointegration of Ti implants could be its coating with bioactive proteins. For bone regeneration, BMP are the most potent osteoinductive proteins and among these, BMP-2 has been reported to induce osteogenic differentiation of stem cells, which can enhance the osseointegration of implants by forming bone at the space between the implants and the implantation site [133,134]. La and collaborators investigated the efficiency of a Ti substrate coated with GO as a delivery carrier for BMP-2, an osteoinductive protein that need to be released over a long time period and Substance P (SP), a stem cells recruiter protein [135], for bone regeneration [136]. In vitro experiments showed that no significant difference was found in SP release between Ti and Ti/GO. Interestingly, the BMP-2 release from the Ti/GO substrate was maintained for 14 days; on the contrary, the release of almost all the BMP-2 content from the Ti substrate occurred on the first day. Furthermore, the bioactivity of the BMP-2 released from the substrate was significantly higher in the GO-coated group. The in vivo study, in which the different substrates were implanted on mouse calvaria, showed that even if no significant difference in bone formation was found between Ti/BMP-2 and Ti/SP/BMP-2, Ti/GO/SP/BMP-2 implants showed much more extensive bone formation than Ti/GO/BMP-2 implants confirming that the presence of GO preserves the bioactivity of proteins (Figure 7).

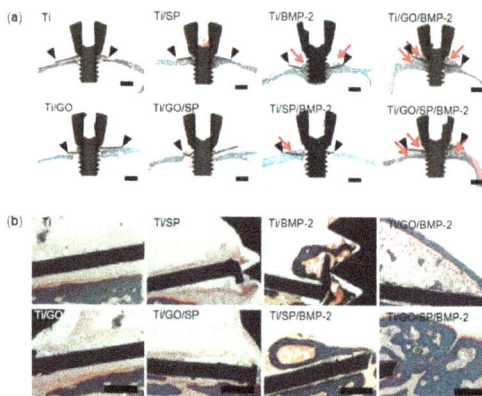

Figure 7. Bone regeneration of Ti implants with or without GO coating and BMP-2/SP loading in mouse calvarial defects 8 weeks after treatment. The red arrowheads indicate the newly formed bone, the black arrowheads indicate the implant at (**a**) 12.5× magnification and (**b**) 100× magnification. Reproduced with permissions from [136].

Implant GO-coating represents an advantage also due to its antibacterial properties. Despite the fixation of graphene on Ti implant makes it to lose a bit of antibacterial activity, there is the possibility to functionalize the coating with antibacterial substances, such as antibiotics or silver nanoparticles. As demonstrated by Quian and colleagues, minocycline hydrochloride added to GO-coating improved the antibacterial activity against aerobic or facultative anaerobic bacteria (*Staphylococcus aureus*), facultative anaerobic bacteria (*E. coli*) and anaerobic bacteria (*S. mutans*) thanks to the synergic effect of GO contact-killing and minocycline release-killing [137]. In another study, the group of Jianfeng tested the antimicrobial activity of GO-silver coating on Ti, showing that it is very prominent against *S. mutans* and *P. gingivalis* [138]. This suggests that this multiphase nanocomposite could be helpful in the prevention of implant-associated infection.

As previously reported, human PDLSCs are promising DSCs that can be used as an alternative to human BM-MSCs and GO is believed to be a useful platform for modulating structure and function of these cells. Nevertheless, the behavior of PDLSCs on GO-coated Ti substrates is not fully understood. In the work of Zhou and colleagues, the morphology, proliferation and osteogenic differentiation potential of PDLSCs seeded on GO-Ti scaffolds were evaluated and compared to those obtained on sodium titanate (Na-Ti) substrates [139]. The results of this study showed that, when the cells were seeded onto GO-coated Ti scaffolds, they exhibited higher proliferation rate, ALP activity and up-regulation of gene expression levels of the osteogenesis-related markers COLI, ALP, Bone Sialoprotein (BSP), RUNX2 and OCN respect to those cultured on the Na-Ti substrate. In addition, GO enhanced the expression of RUNX2, BSP and OCN also at a protein level. The authors concluded that the combination of GO, PDLSCs and Na-Ti can represent a big step forward in regenerative dentistry. Nevertheless, further additional studies will be necessary to elucidate how to use GO as a biocompatible and implantable platform for the delivery of therapeutic proteins for applications related to regenerative medicine, especially for the success of Ti dental implants.

3.2. Graphene and Membranes

To increase efficiency of bone repair, especially in periodontal and periimplant bone defects, the use of Guided Bone Regeneration (GBR) membranes has been important for years in oral surgery. These membranes are placed on the bone defect regeneration site and their principal function is to not allow infiltration of soft tissue cells into the growing bone. Thus, these act as physical barriers that separate connective tissue and the regenerating bone providing the slow osteogenic cells migration [140]. In the dental field, the use of membrane for GBR was introduced in 1980' to stop cell migration from gingival connective tissue and epithelium to periodontal defect [141]. In the last decades, several types of membrane, non-resorbable and bioabsorbable, have been developed and tested for the application in GBR (Table 1) [142].

Table 1. Examples of different types of membrane used for Guided Bone Regeneration (GBR).

Type	Name	Materials	Characteristics
Non-resorbable	Gore-Tex	Expanded polytetrafluoroethylene	Good space maintainer, handling
	Gore-Tex-Ti	Titanium-reinforced polytetrafluoroethylene	Ideal for ridge augmentation and grafting in bone defects
Resorbable natural	Tutodent	Collagen Type I from bovine pericardium	Resorbable rate: 8–16 weeks
	Osseoguard Flex	Collagen Type I and III from bovine dermis	Resorbable rate: 6–9 months
Resorbable synthetic	Epi-Guide	Poly-D,L-lactic acid	Resorbable rate: 6–12 months
	ResolutAdapt	Poly-D,L-lactide/Co-glycolide	Resorbable rate: 5–6 months, good space maintainer

All membranes need to satisfy the five main criteria underlined by Scantlebury: biocompatibility, space-making, cell-occlusiveness, tissue integration and clinical manageability [143]. In order to improve these main criteria, graphene started recently to be used in this field. For example, De Marco and colleagues enriched collagen membranes with two different concentration of GO (2 μg/mL and 10 μg/mL) and tested the effect on Human Gingival Fibroblasts (HGFs) [144]. They showed that the presence of GO on collagen membranes altered membrane features: it conferred a lower deformability, higher stiffness and reduced hydration and increased roughness compared to non-coated membranes. These changes favored the proliferation of HGFs, avoiding any inflammatory response, as demonstrated by the reduction of IL-6 and Prostaglandin E2 (PGE2) secretion after 3 days of culture and facilitated proteins adhesion to the membrane (Figure 8a).

The same GO-coating applied on collagen membranes was tested using DPSCs to analyze osteoblastic differentiation and inflammatory response [145]. The researchers found that GO-coating significantly increased cell viability on a concentration-dependent manner and promoted DPSCs osteoblastic differentiation, as demonstrated by the up-regulation of BMP-2, RUNX2 and OSX gene expression. Moreover, 10 μg/mL GO enrichment reduced the expression of TNF-α at all experimental times compared to the control membrane, whereas 2 μg/mL GO enrichment significantly decreased TNF-α expression only after 21 days of culture. Interestingly, Hematoxylin-Eosin staining revealed that cells were not able to penetrate into the membrane and the more concentrated GO coating led to the formation of a thicker cell layer (Figure 8b). The authors concluded that the GO enrichment of collagen membranes promoted osteoblastic differentiation process, decreased inflammation and was compatible with cell viability in a dose-dependent manner. These new membranes could be considered as a valid alternative to substitute conventional collagen membranes.

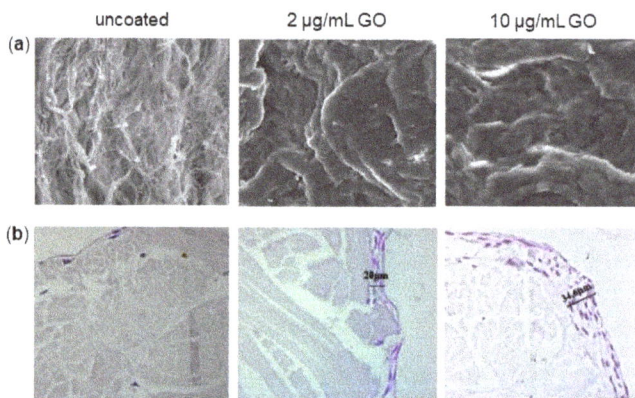

Figure 8. Different GO coating concentration on collagen membrane from porcine dermis. (**a**) SEM images of uncoated, 2 μg/mL and 10 μg/mL GO-coated membranes. 4.05 k magnification. (**b**) Hematoxylin-Eosin staining of uncoated, 2 μg/mL and 10 μg/mL GO-coated membranes with DPSCs after 28 days of culture. 40× magnification. Reproduced with permissions from [144,145].

3.3. Graphene and Resins, Cements and Adhesives

Resins, cements and adhesives are the most used materials for dental restoration. Nevertheless, their porosity and adhesiveness make them receptacles for bacteria in the oral cavity and in particular in proximity of dental restoration. These polymeric materials, indeed, facilitate the adhesion of bacteria and the formation of biofilms that are the main cause of dental restoration failure [146]. An example of the application of graphene to dental commercial materials is reported in the study of Bregnocchi and coworkers [146]. The authors added Graphene Nanoplatelets (GNPs) as nanofiller to a commercial dental adhesive. The generated graphene nanocomposite significantly inhibited the growth of *S. mutans*, without altering the standard adhesion properties of the dental adhesive.

One of the main limitation of GNPs in dental application is its grey color. In order to ameliorate this aspect, Zanni and colleagues tested a hybrid material composed by Zinc Oxide Nanorods (ZnO-NRs) grown on GNPs, thus combining the antimicrobial effect of GNPs with the light color and biocidal properties of ZnO-NRs [147]. The authors studied the effect of Zinc Oxide Nanorods-Decorated Graphene Nanoplatelets (ZNGs) against *S. mutans*. The results showed that up to the 95% *S. mutans* cell viability reduction was obtained using ZNGs even at a very low concentration (5 µg/mL). Moreover, the biomass and exopolysaccharide production, which are necessary for the biofilm formation, were evaluated: the ZNGs showed to behave as an obstacle for the biofilm growth. The authors concluded that the use of ZNGs is a viable instrument to control caries disease as it decreases *S. mutans* growth.

Biris and colleagues successfully prepared graphene sheets embedded with various amounts of gold nanoparticles (Gr-Au-x) using the radiofrequency catalytic chemical vapor deposition technique over a Au_x/MgO catalyst [148]. Sarosi and colleagues analyze the effect of these sheets used as nanofiller for some dental nanocomposites based on BisGMA/triethyleneglicol dimethacrylate matrix [149]. The authors concluded that the graphene–gold nanoparticles could be very promising filler for the dental nanocomposites because the reinforcement with high percent of nanoparticles is a good solution to improve physicochemical properties.

As a last example, Li and collaborators demonstrated that the functionalization of glass ionomer cements with fluorinated graphene is not only useful to inhibit bacterial growth but also to ameliorate mechanical properties of the cements, increasing microhardness and compressive strength and decreasing friction coefficient, all important parameters for cements [150].

3.4. Graphene and Teeth-Whitening

Nowadays, a lot of people would change the color of their teeth, which often is ruined by food, beverage and smoke; consequently, the level of demand for tooth-whitening is increasing. Tooth color is determined by intrinsic factors, associated with the light scattering and absorption properties of the enamel and dentine and extrinsic factors, associated with the absorption of materials (e.g., tea, red wine) on the surface of the enamel.

The products used for tooth-whitening are extensively-based upon Hydrogen Peroxide (H_2O_2) chemistry, because of its bleach-properties and the techniques constantly evolve to ameliorate the rate of whitening, the ease of the treatment and to decrease the side effects. The most common risks of teeth-whitening include increase dental sensitivity and gingival irritation, besides changes in tooth microstructure. These effects are directly related to the concentration of H_2O_2, the duration of the treatment and the non-bleach component of the products [151].

To improve the bleaching process and decrease the time of the treatment, the light-activation of peroxide-based products is used, but novel ways are being found to ameliorate the efficiency of teeth-whitening. Several studies reported that the use of activating agents, such as nanoHAp or Fe(III) phthalocyanine, together with H_2O_2 is useful for the purpose [152]. Based on these evidences, Su and coworkers proposed the application of a nanocomposite made of rGO and Cobalt Tetraphenylporphyrin (CoTPP) as a catalyst for tooth bleaching (Figure 9a) [120]. They then evaluated its efficacy on tooth-whitening compared to the whitelight irradiation and D&C Red 34 or Orange No. 4, two representative stain solutions. They concluded that the use of H_2O_2 plus CoTPP/rGO under photoactivation increased the whitening effect of H_2O_2 and decreased the treatment time (Figure 9b). This result might be explained because the active radicals produced by H_2O_2 have a short life and so their main effect is to penetrate the tooth structure and then start a radical generation mechanism. When photoirradiation is used, radical generation can be started deeper in the structure. Moreover, photoactivating H_2O_2 plus CoTPP/rGO increase the reactions between H_2O_2 and stain molecules.

(a)

(b)

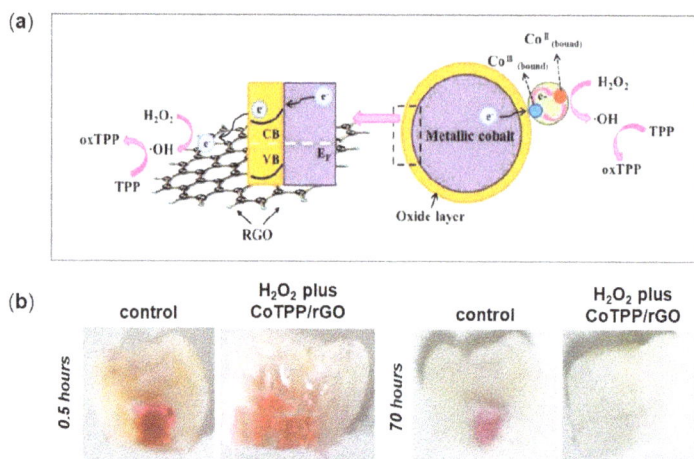

Figure 9. Strategies based on graphene for improving teeth-whitening. (**a**) A schematic diagram illustrating the enhanced peroxidase-like catalytic activity of the rGO-Co. Reactions of Cobalt Tetraphenylporphyrin (CoTPP) with Hydrogen Peroxide (H_2O_2): Co_{III} TPP–$e_2$1(1/2) $H_2O_2 \rightarrow (Co_{IV})$ TPP-OH$_2$ (Co_{IV}) TPP-OH $\rightarrow$ 2 Co_{III} TPP1$O_2$12H$_1$. (**b**) Photographs of teeth stained with dye D&C Red 34 and bleached using H_2O_2 alone or H_2O_2 plus CoTPP/RGO for 0.5 (left) or 70 h (right). Reproduced with permissions of [150].

4. Limitations on the Use of Graphene

While graphene is one of the most promising materials in nanotechnology, the synthesis on large scale represents an important aspect to be evaluated. Most of the studies described so far have been carried out using small amount of uncontaminated and controlled defect-free samples [153], but in the perspective of using graphene for applications on large scale, defects generated during the production process need to be investigated. Various types of defects can generate spontaneously and their formation is difficult to predict; they depend greatly on the method used for the production [154]. Defects change electronic structure, susceptibility and reactivity of graphene and its derivatives [155]. Most of the production techniques lead to the formation of graphene mixture, that differ in size, shape and number of layer and are often contaminated with hydrocarbons or organic molecules which alter the integration and the compatibility with cells and tissues [153]. These evidences represent a substantial limitation on the study and use of graphene and graphene-based nanomaterials.

Another limitation on graphene biomedical application is the scarce existing information on in vivo toxicity mechanisms and more studies are needed to support safe biological applications of these materials. However, some reports showed that graphene-based nanomaterials mainly accumulate in liver, spleen and lung after intravenous administration [43]. Few studies focused on graphene-based nanomaterials elimination half-life indicating that small-sized nanomaterials present fast elimination [45,156], but further work is needed. In addition, because of its recent discovery, little is known about long-term toxicity of graphene and its derivatives and this issue represents a restriction for clinical approaches.

Therefore, the scarce synthetic control, the high variability in production and the multiple parameters that modulate activity and toxicity of graphene and graphene-based nanomaterials, together with the scarce available studies, are big limitations for large scale application of graphene and its derivatives.

5. Conclusions and Future Perspectives

The research on biomedical applications of graphene-based nanomaterials has seen dramatic progress in the last few years. In this review, we focused on the recent advances in dental tissue engineering using graphene and its related nanomaterials. Graphene exhibits numerous outstanding properties. In addition to its exceptional mechanical strength, electrical conductivity and thermal stability, what makes graphene extremely interesting is the possibility to functionalize and combine it with different biomaterials and biomolecules. This confers new properties to existing materials and allows the generation of nanocomposites with enhanced characteristics.

In this review, we discussed recent studies concerning the biocompatibility, cytotoxicity and antibacterial activity of graphene-based nanomaterials both in vitro and in vivo. Then, we analyzed the effect of graphene-based nanomaterials on the adhesion, proliferation and differentiation of stem cells, focusing in particular on stem cells of dental origin. We examined the potential of graphene to promote osteogenic differentiation of stem cells, which is a key point for its future application in dental field. In this context, we presented recent studies employing graphene and its related nanomaterials for surface modification of dental implants or other scaffolds used in dentistry, such as membranes, resins and adhesives. As clearly highlighted in this review, graphene-based nanomaterials have emerged as promising scaffolds for a wide range of biomedical applications and all the presented works seem to agree with the fact that existing dental materials show improved characteristics following the addition of graphene; nevertheless, their behavior is closely dependent on graphene's physicochemical properties, such as surface functionalization, coating and size.

The increased interest in graphene and its derivatives have led to concerns about the risk of exposure not only to humans but also to the environment. Therefore, evaluation of the safety and potential risks of these nanomaterials is mandatory to ensure the safe use of graphene materials in biomedical applications. However, there is still much investigation to perform in order to assess the potential long-term toxicity of graphene and its composites and their effects on different cells, tissues and organs, including those of oral cavity. Moreover, in-depth studies are necessary to understand cell-signaling, metabolic pathways and osteogenic effects triggered by graphene-based nanomaterials. Overall, we believe that the use of graphene-based nanomaterials in the dental field deserves to be deeply explored as it can lead to even more reliable dental treatments in the near future.

Author Contributions: B.Z. conceived, designed the study and critically revised the article. F.S.L., L.S. and L.F. acquired the data. R.G., C.G. and G.B. wrote the paper. A.P., I.A. and E.B. critically revised the article. Final approval was given by all the Authors.

Acknowledgments: The authors would like to acknowledge the contribution of the COST Action CA16122.

Conflicts of Interest: The authors declare no conflict of interest. No grants were awarded in relation to this study.

References

1. Langer, R.; Vacanti, J.P. Tissue engineering. *Science* **1993**, *260*, 920–926. [CrossRef] [PubMed]
2. Kim, B.S.; Mooney, D.J. Development of biocompatible synthetic extracellular matrices for tissue engineering. *Trends Biotechnol.* **1998**, *16*, 224–230. [CrossRef]
3. Hohlfeld, J.; de Buys Roessingh, A.; Hirt-Burri, N.; Chaubert, P.; Gerber, S.; Scaletta, C.; Hohlfeld, P.; Applegate, L.A. Tissue engineered fetal skin constructs for paediatric burns. *Lancet* **2005**, *366*, 840–842. [CrossRef]
4. Cao, Y.; Vacanti, J.P.; Paige, K.T.; Upton, J.; Vacanti, C.A. Transplantation of chondrocytes utilizing a polymer-cell construct to produce tissue-engineered cartilage in the shape of a human ear. *Plast. Reconstr. Surg.* **1997**, *100*, 297–302. [CrossRef] [PubMed]
5. McKeon-Fischer, K.D.; Flagg, D.H.; Freeman, J.W. Coaxial electrospun poly(ε-caprolactone), multiwalled carbon nanotubes, and polyacrylic acid/polyvinyl alcohol scaffold for skeletal muscle tissue engineering. *J. Biomed. Mater. Res. A* **2011**, *99*, 493–499. [CrossRef] [PubMed]

6. Niklason, L.E.; Gao, J.; Abbott, W.M.; Hirschi, K.K.; Houser, S.; Marini, R.; Langer, R. Functional arteries grown in vitro. *Science* **1999**, *284*, 489–493. [CrossRef] [PubMed]

7. Warnke, P.H.; Springer, I.N.; Wiltfang, J.; Acil, Y.; Eufinger, H.; Wehmoller, M.; Russo, P.A.; Bolte, H.; Sherry, E.; Behrens, E.; et al. Growth and transplantation of a custom vascularised bone graft in a man. *Lancet* **2004**, *364*, 766–770. [CrossRef]

8. Abou Neel, E.A.; Chrzanowski, W.; Salih, V.M.; Kim, H.W.; Knowles, J.C. Tissue engineering in dentistry. *J. Dent.* **2014**, *42*, 915–928. [CrossRef] [PubMed]

9. Gardin, C.; Ricci, S.; Ferroni, L. Dental Stem Cells (DSCs): Classification and Properties. In *Dental Stem Cells: Regenerative Potential*; Zavan, B., Bressan, E., Eds.; Springer International Publishing AG: Basel, Switzerland, 2016; pp. 1–25.

10. Malhotra, N.; Mala, K. Regenerative endodontics as a tissue engineering approach: Past, current and future. *Aust. Endod. J.* **2012**, *38*, 137–148. [CrossRef] [PubMed]

11. Cordeiro, M.M.; Dong, Z.; Kaneko, T.; Zhang, Z.; Miyazawa, M.; Shi, S.; Smith, A.J.; Nor, J.E. Dental pulp tissue engineering with stem cells from exfoliated deciduous teeth. *J. Endod.* **2008**, *34*, 962–969. [CrossRef] [PubMed]

12. Nakahara, T.; Nakamura, T.; Kobayashi, E.; Kuremoto, K.; Matsuno, T.; Tabata, Y.; Eto, K.; Shimizu, Y. In situ tissue engineering of periodontal tissues by seeding with periodontal ligament derived cells. *Tissue Eng.* **2004**, *10*, 537–544. [CrossRef] [PubMed]

13. Hu, B.; Nadiri, A.; Kuchler-Bopp, S.; Perrin-Schmitt, F.; Peters, H.; Lesot, H. Tissue engineering of tooth crown, root, and periodontium. *Tissue Eng.* **2006**, *12*, 2069–2075. [CrossRef] [PubMed]

14. Sakai, V.T.; Zhang, Z.; Dong, Z.; Neiva, K.G.; Machado, M.A.; Shi, S.; Santos, C.F.; Nör, J.E. SHED differentiate into functional odontoblasts and endothelium. *J. Dent. Res.* **2010**, *89*, 791–796. [CrossRef] [PubMed]

15. Chen, H.; Tang, Z.; Liu, J.; Sun, K.; Chang, S.R.; Peters, M.C.; Mansfield, J.F.; Czajka-Jakubowska, A.; Clarkson, B.H. Acellular synthesis of a humamn enamel-like microstructure. *Adv. Mater.* **2006**, *18*, 1846–1851. [CrossRef]

16. Duailibi, M.T.; Duailibi, S.E.; Young, C.S.; Bartlett, J.D.; Vacanti, J.P.; Yelick, P.C. Bioengineered teeth from cultured rat tooth bud cells. *J. Dent. Res.* **2004**, *83*, 523–528. [CrossRef] [PubMed]

17. Ikeda, E.; Morita, R.; Nakao, K.; Ishida, K.; Nakamura, T.; Takano-Yamamoto, T.; Ogawa, M.; Mizuno, M.; Kasugai, S.; Tsuji, T. Fully functional bioengineered tooth replacement as an organ replacement therapy. *Proc. Natl. Acad. Sci. USA* **2009**, *106*, 13475–13480. [CrossRef] [PubMed]

18. Sumita, Y.; Honda, M.J.; Ohara, T.; Tsuchiya, S.; Sagara, H.; Kagami, H.; Ueda, M. Performance of collagen sponge as a 3-D scaffold for tooth-tissue engineering. *Biomaterials* **2006**, *27*, 3238–3248. [CrossRef] [PubMed]

19. Kuo, T.F.; Huang, A.T.; Chang, H.H.; Lin, F.H.; Chen, S.T.; Chen, R.S.; Chou, C.H.; Lin, H.C.; Chiang, H.; Chen, M.H. Regeneration of dentin-pulp complex with cementum and periodontal ligament formation using dental bud cells in gelatin-chondroitin-hyaluronan tri-copolymer scaffold in swine. *J. Biomed. Mater. Res. A* **2008**, *86*, 1062–1068. [CrossRef] [PubMed]

20. Kirkham, J.; Firth, A.; Vernals, D.; Boden, N.; Robinson, C.; Shore, R.C.; Brookes, S.J.; Aggeli, A. Self-assembling peptide scaffolds promote enamel remineralization. *J. Dent. Res.* **2007**, *86*, 426–430. [CrossRef] [PubMed]

21. Xu, W.P.; Zhang, W.; Asrican, R.; Kim, H.J.; Kaplan, D.L.; Yelick, P.C. Accurately shaped tooth bud cell-derived mineralized tissue formation on silk scaffolds. *Tissue Eng. Part A* **2008**, *14*, 549–557. [CrossRef] [PubMed]

22. Nishida, E.; Miyaji, H.; Kato, A.; Takita, H.; Iwanaga, T.; Momose, T.; Ogawa, K.; Murakami, S.; Sugaya, T.; Kawanami, M. Graphene oxide scaffold accelerates cellular proliferative response and alveolar bone healing of tooth extraction socket. *Int. J. Nanomed.* **2016**, *11*, 2265–2277.

23. Goenka, S.; Sant, V.; Sant, S. Graphene-based nanomaterials for drug delivery and tissue engineering. *J. Controll. Release* **2014**, *173*, 75–88. [CrossRef] [PubMed]

24. Shin, S.R.; Li, Y.C.; Jang, H.L.; Khoshakhlagh, P.; Akbari, M.; Nasajpour, A.; Zhang, Y.S.; Tamayol, A.; Khademhosseini, A. Graphene-based materials for tissue engineering. *Adv. Drug. Deliv. Rev.* **2016**, *105*, 255–274. [CrossRef] [PubMed]

25. Jastrzębska, A.M.; Kurtycz, P.; Olszyna, A.R. Recent advances in graphene family materials toxicity investigations. *J. Nanopart. Res.* **2012**, *14*, 1320–1340. [CrossRef] [PubMed]

26. Zhao, H.; Ding, R.; Zhao, X.; Li, Y.; Qu, L.; Pei, H.; Yildirimer, L.; Wu, Z.; Zhang, W. Graphene-based nanomaterials for drug and/or gene delivery, bioimaging, and tissue engineering. *Drug Discov. Today* **2017**, *22*, 1302–1317. [CrossRef] [PubMed]

27. Novoselov, K.S.; Geim, A.K.; Morozov, S.V.; Jiang, D.; Zhang, Y.; Dubonos, S.V.; Grigorieva, I.V.; Firsov, A.A. Electric field effect in atomically thin carbon films. *Science* **2004**, *306*, 666–669. [CrossRef] [PubMed]

28. Liu, Z.; Robinson, J.T.; Sun, X.; Dai, H. PEGylated nanographene oxide for delivery of water-insoluble cancer drugs. *J. Am. Chem. Soc.* **2008**, *130*, 10876–10877. [CrossRef] [PubMed]

29. Wang, Y.B.; Li, H.F.; Cheng, Y.; Zheng, Y.F.; Ruan, L.Q. In vitro and in vivo studies on Ti-based bulk metallic glass as potential dental implant material. *Mater. Sci. Eng. C Mater. Biol. Appl.* **2013**, *33*, 3489–3497. [CrossRef] [PubMed]

30. Chan, B.P.; Leong, K.W. Scaffolding in tissue engineering: General approaches and tissue-specific considerations. *Eur. Spine J.* **2008**, *17*, 467–479. [CrossRef] [PubMed]

31. Lee, W.C.; Loh, K.P.; Lim, C.T. When stem cells meet graphene: Opportunities and challenges in regenerative medicine. *Biomaterial* **2018**, *155*, 236–250.

32. Zhang, Y.; Ali, S.F.; Dervishi, E.; Xu, Y.; Li, Z.; Casciano, D.; Biris, A.S. Cytotoxicity effects of graphene and single-wall carbon nanotubes in neural phaeochromocytoma-derived PC12 cells. *ACS Nano* **2010**, *4*, 3181–3186. [CrossRef] [PubMed]

33. Chang, Y.; Yang, S.T.; Liu, J.H.; Dong, E.; Wang, Y.; Cao, A.; Liu, Y.; Wang, H. In vitro toxicity evaluation of graphene oxide on A549 cells. *Toxicol. Lett.* **2011**, *200*, 201–210. [CrossRef] [PubMed]

34. Van Tienhoven, E.A.; Korbee, D.; Schipper, L.; Verharen, H.W.; De Jong, W.H. In vitro and in vivo (cyto)toxicity assays using PVC and LDPE as model materials. *J. Biomed. Mater. Res. A* **2006**, *78*, 175–182. [CrossRef] [PubMed]

35. Peng, L.; Gao, Y.; Xue, Y.N.; Huang, S.W.; Zhuo, R.X. Cytotoxicity and in vivo tissue compatibility of poly(amidoamine) with pendant aminobutyl group as a gene delivery vector. *Biomaterials* **2010**, *31*, 4467–4476. [CrossRef] [PubMed]

36. Compton, O.C.; Nguyen, S.T. Graphene oxide, highly reduced graphene oxide, and graphene: Versatile building blocks for carbon-based materials. *Small* **2010**, *6*, 711–723. [CrossRef] [PubMed]

37. Sasidharan, A.; Panchakarla, L.S.; Chandran, P.; Menon, D.; Nair, S.; Rao, C.N.; Koyakutty, M. Differential nano-bio interactions and toxicity effects of pristine versus functionalized graphene. *Nanoscale* **2011**, *3*, 2461–2464. [CrossRef] [PubMed]

38. Das, S.; Singh, S.; Singh, V.; Joung, D.; Dowding, J.M.; Anderson, D.R.J.; Zhai, L.; Khondaker, S.I.; Self, W.T.; Seal, S. Oxygenated Functional Group Density on Graphene Oxide: Its Effect on Cell Toxicity. *Part. Part. Syst. Characterization* **2013**, *30*, 148–157. [CrossRef]

39. Bagri, A.; Mattevi, C.; Acik, M.; Chabal, Y.J.; Chhowalla, M.; Shenoy, V.B. Structural evolution during the reduction of chemically derived graphene oxide. *Nat. Chem.* **2010**, *2*, 581–587. [CrossRef] [PubMed]

40. Zhang, W.; Yan, L.; Li, M.; Zhao, R.; Yang, X.; Ji, T.; Gu, Z.; Yin, J.J.; Gao, X.; Nie, G. Deciphering the underlying mechanisms of oxidation-state dependent cytotoxicity of graphene oxide on mammalian cells. *Toxicol. Lett.* **2015**, *237*, 61–71. [CrossRef] [PubMed]

41. Schieber, M.; Chandel, N.S. ROS function in redox signaling and oxidative stress. *Curr. Biol.* **2014**, *24*, R453–R462. [CrossRef] [PubMed]

42. Duan, G.; Zhang, Y.; Luan, B.; Weber, J.K.; Zhou, R.W.; Yang, Z.; Zhao, L.; Xu, J.; Luo, J.; Zhou, R. Graphene-Induced Pore Formation on Cell Membranes. *Sci. Rep.* **2017**, *7*, 42767. [CrossRef] [PubMed]

43. Pinto, A.M.; Gonçalves, I.C.; Magalhães, F.D. Graphene-based materials biocompatibility: A review. *Colloids Surf. B Biointerfaces* **2013**, *111*, 188–202. [CrossRef] [PubMed]

44. Wang, K.; Ruan, J.; Song, H.; Zhang, J.; Wo, Y.; Guo, S.; Cui, D. Biocompatibility of Graphene Oxide. *Nanoscale Res. Lett.* **2011**, *6*, 8. [CrossRef] [PubMed]

45. Duch, M.C.; Budinger, G.R.; Liang, Y.T.; Soberanes, S.; Urich, D.; Chiarella, S.E.; Campochiaro, L.A.; Gonzalez, A.; Chandel, N.S.; Hersam, M.C.; et al. Minimizing oxidation and stable nanoscale dispersion improves the biocompatibility of graphene in the lung. *Nano Lett.* **2011**, *11*, 5201–5207. [CrossRef] [PubMed]

46. Yang, K.; Gong, H.; Shi, X.; Wan, J.; Zhang, Y.; Liu, Z. In vivo biodistribution and toxicology of functionalized nano-graphene oxide in mice after oral and intraperitoneal administration. *Biomaterials* **2013**, *34*, 2787–2795. [CrossRef] [PubMed]

47. Nayak, T.R.; Andersen, H.; Makam, V.S.; Khaw, C.; Bae, S.; Xu, X.; Ee, P.L.; Ahn, J.H.; Hong, B.H.; Pastorin, G.; Özyilmaz, B. Graphene for controlled and accelerated osteogenic differentiation of human mesenchymal stem cells. *ACS Nano* **2011**, *5*, 4670–4678. [CrossRef] [PubMed]

48. Ankrum, J.; Karp, J.M. Mesenchymal stem cell therapy: Two steps forward, one step back. *Trends Mol. Med.* **2010**, *16*, 203–209. [CrossRef] [PubMed]

49. Lee, W.C.; Lim, C.H.Y.; Shi, H.; Tang, L.A.; Wang, Y.; Lim, C.T.; Loh, K.P. Origin of enhanced stem cell growth and differentiation on graphene and graphene oxide. *ACS Nano* **2011**, *5*, 7334–7341. [CrossRef] [PubMed]

50. Wei, C.; Liu, Z.; Jiang, F.; Zeng, B.; Huang, M.; Yu, D. Cellular behaviours of bone marrow-derived mesenchymal stem cells towards pristine graphene oxide nanosheets. *Cell Prolif.* **2017**, *50*. [CrossRef] [PubMed]

51. Crowder, S.W.; Prasai, D.; Rath, R.; Balikov, D.A.; Bae, H.; Bolotin, K.I.; Sung, H.J. Three-dimensional graphene foams promote osteogenic differentiation of human mesenchymal stem cells. *Nanoscale* **2013**, *5*, 4171–4176. [CrossRef] [PubMed]

52. Depan, D.; Girase, B.; Shah, J.S.; Misra, R. Structure-process-property relationship of the polar graphene oxide-mediated cellular response and stimulated growth of osteoblasts on hybrid chitosan network structure nanocomposite scaffolds. *Acta Biomater.* **2011**, *7*, 3432–3445. [CrossRef] [PubMed]

53. Lee, J.H.; Shin, Y.C.; Jin, O.S.; Kang, S.H.; Hwang, Y.S.; Park, J.C.; Hong, S.W.; Han, D.W. Reduced graphene oxide-coated hydroxyapatite composites stimulate spontaneous osteogenic differentiation of human mesenchymal stem cells. *Nanoscale* **2015**, *7*, 11642–11651. [CrossRef] [PubMed]

54. Fratzl, P.; Weinkanwer, R. Nature's hierarchical materials. *Prog. Mater. Sci.* **2007**, *52*, 1263–1334. [CrossRef]

55. Geiger, M.; Li, R.H.; Friess, W. Collagen sponges for bone regeneration with rhBMP-2. *Adv. Drug Deliv. Rev.* **2003**, *55*, 1613–1629. [CrossRef] [PubMed]

56. Kang, S.; Park, J.B.; Lee, T.J.; Ryud, S.; Bhange, S.H.; Laf, W.G.; Noh, M.K.; Hong, B.H.; Kim, B.S. Covalent conjugation of mechanically stiff graphene oxide flakes to threedimensional collagen scaffolds for osteogenic differentiation of human mesenchymal stem cells. *Carbon* **2015**, *83*, 162–172. [CrossRef]

57. Nishida, E.; Miyaji, H.; Takita, H.; Kanayama, I.; Tsuji, M.; Akasaka, T.; Sugaya, T.; Sakagami, R.; Kawanami, M. Graphene oxide coating facilitates the bioactivity of scaffold material for tissue engineering. *Jpn. J. Appl. Phys.* **2014**, *53*, 6JD04. [CrossRef]

58. Kim, T.G.; Park, T.G. Biomimicking extracellular matrix: Cell adhesive RGD peptide modified electrospun poly (D, L-lactic-co-glycolic acid) nanofiber mesh. *Tissue Eng.* **2006**, *12*, 221–233. [CrossRef] [PubMed]

59. Kim, T.H.; Lee, D.; Choi, J.W. Live cell biosensing platforms using graphene-based hybrid nanomaterials. *Biosens. Bioelectron.* **2017**, *94*, 485–499. [CrossRef] [PubMed]

60. Kang, E.S.; Kim, D.S.; Han, Y.; Son, H.; Chung, Y.H.; Min, J.; Kim, T.H. Three-Dimensional Graphene-RGD Peptide Nanoisland Composites That Enhance the Osteogenesis of Human Adipose-Derived Mesenchymal Stem Cells. *Int. J. Mol. Sci.* **2018**, *19*, 669. [CrossRef] [PubMed]

61. Casadei, A.; Epis, R.; Ferroni, L.; Tocco, I.; Gardin, C.; Bressan, E.; Sivolella, S.; Vindigni, V.; Pinton, P.; Mucci, G.; et al. Adipose tissue regeneration: A state of the art. *J. Biomed. Biotechnol.* **2012**, *2012*, 462543. [CrossRef] [PubMed]

62. Lim, K.T.; Seonwoo, H.; Choi, K.S.; Jin, H.; Jang, K.J.; Kim, J.; Kim, J.W.; Kim, S.Y.; Choung, P.H.; Chung, J.H. Pulsed-Electromagnetic-Field-Assisted Reduced Graphene Oxide Substrates for Multidifferentiation of Human Mesenchymal Stem Cells. *Adv. Healthc. Mater.* **2016**, *5*, 2069–2079. [CrossRef] [PubMed]

63. Ferroni, L.; Tocco, I.; De Pieri, A.; Menarin, M.; Fermi, E.; Piattelli, A.; Gardin, C.; Zavan, B. Pulsed magnetic therapy increases osteogenic differentiation of mesenchymal stem cells only if they are pre-committed. *Life Sci.* **2016**, *152*, 44–51. [CrossRef] [PubMed]

64. Gronthos, S.; Mankani, M.; Brahim, J.; Robey, P.G.; Shi, S. Postnatal human dental pulp stem cells (DPSCs) in vitro and in vivo. *Proc. Natl. Acad. Sci. USA* **2000**, *97*, 13625–13630. [CrossRef] [PubMed]

65. Shi, S.; Robey, P.G.; Gronthos, S. Comparison of human dental pulp and bone marrow stromal stem cells by cDNA microarray analysis. *Bone* **2001**, *29*, 532–539. [CrossRef]

66. D'Aquino, R.; Papaccio, G.; Laino, G.; Graziano, A. Dental pulp stem cells: A promising tool for bone regeneration. *Stem Cell Rev.* **2008**, *4*, 21–26. [CrossRef] [PubMed]

67. Rosa, V.; Xie, H.; Dubey, N.; Madanagopal, T.T.; Rajan, S.S.; Morin, J.L.; Islam, I.; Castro Neto, A.H. Graphene oxide-based substrate: Physical and surface characterization, cytocompatibility and differentiation potential of dental pulp stem cells. *Dent. Mater.* **2016**, *32*, 1019–1025. [CrossRef] [PubMed]

68. Saadi, I.; Das, P.; Zhao, M.; Raj, L.; Ruspita, I.; Xia, Y.; Papaioannou, V.E.; Bei, M. Msx1 andTbx2 antagonistically regulate Bmp4 expression during thebud-to-cap stage transition in tooth development. *Development* **2013**, *140*, 2697–2702. [CrossRef] [PubMed]

69. Wang, Y.; Groppe, J.C.; Wu, J.; Ogawa, T.; Mues, G.; D'Souza, R.N.; Kapadia, H. Pathogenic mechanisms of tooth agenesis linked to paired domain mutations in human PAX9. *Hum. Mol. Genet.* **2009**, *18*, 2863–2874. [CrossRef] [PubMed]

70. Ryoo, H.M.; Hoffmann, H.M.; Beumer, T.; Frenkel, B.; Towler, D.A.; Stein, G.S.; Stein, J.L.; van Wijnen, A.J.; Lian, J.B. Stage-specific expression of Dlx-5 during osteoblast differentiation: Involvement in regulation of osteocalcin gene expression. *Mol. Endocrinol.* **1997**, *11*, 1681–1694. [CrossRef] [PubMed]

71. Ye, L.; MacDougall, M.; Zhang, S.; Xie, Y.; Zhang, J.; Li, Z.; Lu, Y.; Mishina, Y.; Feng, J.Q. Deletion of dentin matrix protein-1 leads to a partial failure of maturation of predentin into dentin, hypomineralization, and expanded cavities of pulp and root canal during postnatal tooth development. *J. Biol. Chem.* **2004**, *279*, 19141–19148. [CrossRef] [PubMed]

72. Feng, J.Q.; Luan, X.; Wallace, J.; Jing, D.; Ohshima, T.; Kulkarni, A.B.; D'Souza, R.N.; Kozak, C.A.; MacDougall, M. Genomic organization, chromosomal mapping, and promoter analysis of the mouse dentin sialophosphoprotein (Dspp) gene, which codes for both dentin sialoprotein and dentin phosphoprotein. *J. Biol. Chem.* **1998**, *273*, 9457–9464. [CrossRef] [PubMed]

73. Xie, H.; Chua, M.; Islam, I.; Bentini, R.; Cao, T.; Viana-Gomes, J.C.; Castro Neto, A.H.; Rosa, V. CVD-grown monolayer graphene induces osteogenic but not odontoblastic differentiation of dental pulp stem cells. *Dent. Mater.* **2017**, *33*, e13–e21. [CrossRef] [PubMed]

74. Lee, C.; Wei, X.; Kysar, J.W.; Hone, J. Measurement of the elastic properties and intrinsic strength of monolayer graphene. *Science* **2008**, *321*, 385–388. [CrossRef] [PubMed]

75. Lu, Q.; Pandya, M.; Rufaihah, A.J.; Rosa, V.; Tong, H.J.; Seliktar, D.; Toh, W.S. Modulation of dental pulp stem cell odontogenesis in a tunable PEG-fibrinogen hydrogel system. *Stem Cells Int.* **2015**, *2015*, 525367. [CrossRef] [PubMed]

76. Seo, B.M.; Miura, M.; Gronthos, S.; Bartold, P.M.; Batouli, S.; Brahim, J.; Young, M.; Robey, P.G.; Wang, C.Y.; Shi, S. Investigation of multipotent postnatal stem cells from human periodontal ligament. *Lancet* **2004**, *364*, 149–155. [CrossRef]

77. Lee, M.H.; Kang, J.H.; Lee, S.W. The effect of surface microgrooves and anodic oxidation on the surface characteristics of titanium and the osteogenic activity of human periodontal ligament cells. *Arch. Oral Biol.* **2013**, *58*, 59–66. [CrossRef] [PubMed]

78. Heo, Y.Y.; Um, S.; Kim, S.K.; Park, J.M.; Seo, B.M. Responses of periodontal ligament stem cells on various titanium surfaces. *Oral Dis.* **2011**, *17*, 320–327. [CrossRef] [PubMed]

79. Rodríguez-Lozano, F.J.; García-Bernal, D.; Aznar-Cervantes, S.; Ros-Roca, M.A.; Algueró, M.C.; Atucha, N.M.; Lozano-García, A.A.; Moraleda, J.M.; Cenis, J.L. Effects of composite films of silk fibroin and graphene oxide on the proliferation, cell viability and mesenchymal phenotype of periodontal ligament stem cells. *J. Mater. Sci. Mater. Med.* **2014**, *25*, 2731–2741. [CrossRef] [PubMed]

80. Hu, K.; Gupta, M.K.; Kulkarni, D.D.; Tsukruk, V.V. Ultra-robust graphene oxide-silk fibroin nanocomposite membranes. *Adv. Mater.* **2013**, *25*, 2301–2307. [CrossRef] [PubMed]

81. Park, J.; Park, S.; Ryu, S.; Bhang, S.H.; Kim, J.; Yoon, J.K.; Park, Y.H.; Cho, S.P.; Lee, S.; Hong, B.H.; et al. Graphene-regulated cardiomyogenic differentiation process of mesenchymal stem cells by enhancing the expression of extracellular matrix proteins and cell signaling molecules. *Adv. Healthc. Mater.* **2014**, *3*, 176–181. [CrossRef] [PubMed]

82. Vera-Sánchez, M.; Aznar-Cervantes, S.; Jover, E.; García-Bernal, D.; Oñate-Sánchez, R.E.; Hernández-Romero, D.; Moraleda, J.M.; Collado-González, M.; Rodríguez-Lozano, F.J.; Cenis, J.L. Silk-Fibroin and Graphene Oxide Composites Promote Human Periodontal Ligament Stem Cell Spontaneous Differentiation into Osteo/Cementoblast-Like Cells. *Stem Cells Dev.* **2016**, *25*, 1742–1754. [CrossRef] [PubMed]

83. Xie, H.; Cao, T.; Viana Gomes, J.; Castro Neto, A.H.; Rosa, V. Two and three-dimensional graphene substrates to magnify osteogenic differentiation of periodontal ligament stem cells. *Carbon* **2005**, *93*, 266–275. [CrossRef]

84. Jin, H.J.; Park, S.K.; Oh, W.; Yang, Y.S.; Kim, S.W.; Choi, S.J. Down-regulation of CD105 is associated with multilineage differentiation in human umbilical cord bloodderived mesenchymal stem cells. *Biochem. Biophys. Res. Commun.* **2009**, *381*, 676–681. [CrossRef] [PubMed]

85. Wiesmann, A.; Bühring, H.J.; Mentrup, C.; Wiesmann, H.P. Decreased CD90 expression in human mesenchymal stem cells by applying mechanical stimulation. *Head Face Med.* **2006**, *2*, 8. [CrossRef] [PubMed]

86. Torii, D.; Tsutsui, T.W.; Watanabe, N.; Konishi, K. Bone morphogenetic protein 7 induces cementogenic differentiation of human periodontal ligament-derived mesenchymal stem cells. *Odontology* **2016**, *104*, 1–9. [CrossRef] [PubMed]

87. Wise, G.E.; Frazier-Bowers, S.; D'Souza, R.N. Cellular, molecular, and genetic determinants of tooth eruption. *Crit. Rev. Oral Biol. Med.* **2002**, *13*, 323–334. [CrossRef] [PubMed]

88. Morsczeck, C.; Götz, W.; Schierholz, J.; Zeilhofer, F.; Kühn, U.; Möhl, C.; Sippel, C.; Hoffmann, K.H. Isolation of precursor cells (PCs) from human dental follicle of wisdom teeth. *Matrix Biol.* **2005**, *24*, 155–165. [CrossRef] [PubMed]

89. Olteanu, D.; Filip, A.; Socaci, C.; Biris, A.R.; Filip, X.; Coros, M.; Rosu, M.C.; Pogacean, F.; Alb, C.; Baldea, I.; et al. Cytotoxicity assessment of graphene-based nanomaterials on human dental follicle stem cells. *Colloids Surf. B Biointerfaces* **2015**, *136*, 791–798. [CrossRef] [PubMed]

90. Orecchioni, M.; Jasim, D.A.; Pescatori, M.; Manetti, R.; Fozza, C.; Sgarrella, F.; Bedognetti, D.; Bianco, A.; Kostarelos, K.; Delogu, L.G. Molecular and genomic impact of large and small lateral dimension graphene oxide sheets on human immune cells from healthy donors. *Adv. Healthc. Mater.* **2016**, *2*, 276–287. [CrossRef] [PubMed]

91. Ding, Z.; Zhang, Z.; Ma, H.; Chen, Y. In Vitro Hemocompatibility and Toxic Mechanism of Graphene Oxide on Human Peripheral Blood T Lymphocytes and Serum Albumin. *ACS Appl. Mater. Interfaces* **2014**, *6*, 19797–19807. [CrossRef] [PubMed]

92. Zhi, X.; Fang, H.; Bao, C.; Shen, G.; Zhang, J.; Wang, K.; Guo, S.; Wan, T.; Cui, D. The immunotoxicity of graphene oxides and the effect of PVP-coating. *Biomaterials* **2013**, *34*, 5254–5261. [CrossRef] [PubMed]

93. Orecchioni, M.; Bedognetti, D.; Sgarrella, F.; Marincola, F.M.; Bianco, A.; Delogu, L.G. Impact of carbon nanotubes and graphene on immune cells. *J. Transl. Med.* **2014**, *12*, 138. [CrossRef] [PubMed]

94. Yue, H.; Wei, W.; Yue, Z. The role of the lateral dimension of graphene oxide in the regulation of cellular responses. *Biomaterials* **2012**, *33*, 4013–4021. [CrossRef] [PubMed]

95. Russier, J.; Treossi, E.; Scarsi, A.; Perrozzi, F.; Dumortier, H.; Ottaviano, L.; Meneghetti, M.; Palermo, V.; Bianco, A. Evidencing the mask effect of graphene oxide: A comparative study on primary human and murine phagocytic cells. *Nanoscale* **2013**, *5*, 11234–11247. [CrossRef] [PubMed]

96. Chen, G.Y.; Yang, H.; Lu, C.H.; Chao, Y.C.; Hwang, S.M.; Chen, C.L.; Lo, K.W.; Sung, L.Y.; Luo, W.Y.; Tuan, H.Y.; et al. Simultaneous induction of autophagy and toll-like receptor signaling pathways by graphene oxide. *Biomaterials* **2012**, *33*, 6559–6569. [CrossRef] [PubMed]

97. Kawai, T.; Akira, S. Signaling to NF-kappaB by Toll-like receptors. *Trends Mol. Med.* **2007**, *13*, 460–469. [CrossRef] [PubMed]

98. Kawai, T.; Akira, S. The role of pattern-recognition receptors in innate immunity: Update on Toll-like receptors. *Nat. Immunol.* **2010**, *11*, 373–384. [CrossRef] [PubMed]

99. Dudek, I.; Skoda, M.; Jarosz, A.; Szukiewicz, D. The Molecular Influence of Graphene and Graphene Oxide on the Immune System Under In Vitro and In Vivo Conditions. *Arch. Immunol. Ther. Exp.* **2016**, *64*, 195–215. [CrossRef] [PubMed]

100. Zhou, H.; Zhao, K.; Li, W.; Yang, N.; Liu, Y.; Chen, C.; Wei, T. The interactions between pristine graphene and macrophages and the production of cytokines/chemokines via TLR- and NF-kB-related signaling pathways. *Biomaterials* **2012**, *33*, 6933–6942. [CrossRef] [PubMed]

101. Beek, J.J.; Wimmers, F.; Hato, S.V.; De Vries, I.J.; Sköld, A.E. Dendritic cell cross talk with innate and innate-like effector cells in antitumor immunity: Implications for DC vaccination. *Crit. Rev. Immunol.* **2014**, *34*, 517–536. [CrossRef] [PubMed]

102. Tkach, A.V.; Yanamala, N.; Stanley, S.; Shurin, M.R.; Shurin, G.V.; Kisin, E.R.; Murray, A.R.; Pareso, S.; Khaliullin, T.; Kotchey, G.P.; et al. Graphene oxide, but not fullerenes, targets immunoproteasomes and suppresses antigen presentation by dendritic cells. *Small* **2013**, *9*, 1686–1690. [CrossRef] [PubMed]

103. Khan, S.T.; Musarrat, J.; Al-Khedhairy, A.A. Countering drug resistance, infectious diseases, and sepsis using metal and metal oxides nanoparticles: Current status. *Colloids Surf. B* **2016**, *146*, 70–83. [CrossRef] [PubMed]

104. Olivi, M.; Zanni, E.; De Bellis, G.; Talora, C.; Sarto, M.S.; Palleschi, C.; Flahaut, E.; Monthioux, M.; Rapino, S.; Uccelletti, D.; et al. Inhibition of microbial growth by carbon nanotube networks. *Nanoscale* **2013**, *5*, 9023–9029. [CrossRef] [PubMed]

105. Hu, W.; Peng, C.; Luo, W.; Lv, M.; Li, X.; Li, D.; Huang, Q.; Fan, C. Graphene-based antibacterial paper. *ACS Nano* **2010**, *4*, 4317–4423. [CrossRef] [PubMed]

106. Liu, S.; Zeng, T.H.; Hofmann, M.; Burcombe, E.; Wei, J.; Jiang, R.; Kong, J.; Chen, Y. Antibacterial activity of graphite, graphite oxide, graphene oxide, and reduced graphene oxide: Membrane and oxidative stress. *ACS Nano* **2011**, *5*, 6971–6980. [CrossRef] [PubMed]

107. Pham, V.T.H.; Truong, V.K.; Quinn, M.D.J.; Notley, S.M.; Guo, Y.; Baulin, V.A.; Al Kobaisi, M.; Crawford, R.J.; Ivanova, E.P. Graphene induces formation of pores that kill spherical and rod-shaped bacteria. *ACS Nano* **2015**, *9*, 8458–8467. [CrossRef] [PubMed]

108. Akhavan, O.; Ghaderi, E.; Esfandiar, A. Wrapping bacteria by graphene nanosheets for isolation from environment, reactivation by sonication, and inactivation by near-infrared irradiation. *J. Phys. Chem. B* **2011**, *115*, 6279–6288. [CrossRef] [PubMed]

109. Zhou, R.; Gao, H. Cytotoxicity of graphene: Recent advances and future perspective. *Wiley Interdiscip. Rev. Nanomed. Nanobiotechnol.* **2014**, *6*, 452–474. [CrossRef] [PubMed]

110. Li, J.; Wang, G.; Zhu, H.; Zhang, M.; Zheng, X.; Di, Z.; Liu, X.; Wang, X. Antibacterial activity of large-area monolayer graphene film manipulated by charge transfer. *Sci. Rep.* **2014**, *4*, 4359. [CrossRef] [PubMed]

111. Al-Jumaili, A.; Alancherry, S.; Bazaka, K.; Jacob, M.V. Review on the Antimicrobial Properties of Carbon Nanostructures. *Materials* **2017**, *10*, 1066. [CrossRef] [PubMed]

112. Pang, L.; Dai, C.; Bi, L.; Guo, Z.; Fan, J. Biosafety and Antibacterial Ability of Graphene and Graphene Oxide In Vitro and In Vivo. *Nanoscale Res. Lett.* **2017**, *12*, 564. [CrossRef] [PubMed]

113. Liu, S.; Hu, M.; Zeng, T.H.; Wu, R.; Jiang, R.; Wei, J.; Wang, L.; Kong, J.; Chen, Y. Lateral Dimension-Dependent Antibacterial Activity of Graphene Oxide Sheets. *Langmuir* **2012**, *28*, 12364–12372. [CrossRef] [PubMed]

114. He, J.; Zhu, X.; Qi, Z.; Wang, C.; Mao, X.; Zhu, C.; He, Z.; Li, M.; Tang, Z. Killing Dental Pathogens Using Antibacterial Graphene Oxide. *ACS Appl. Mater. Interfaces* **2015**, *7*, 5605–5611. [CrossRef] [PubMed]

115. Wade, W.G. The Oral Microbiome in Health and Disease. *Pharmacol. Res.* **2013**, *69*, 137–143. [CrossRef] [PubMed]

116. Loesche, W.J. Role of Streptococcus mutans in Human Dental Decay. *Microbiol. Rev.* **1986**, *50*, 353. [PubMed]

117. Ximénez-Fyvie, L.A.; Haffajee, A.D.; Socransky, S.S. Comparison of the Microbiota of Supra-and Subgingival Plaque in Health and Periodontitis. *J. Clin. Periodontol.* **2000**, *27*, 648–657. [CrossRef] [PubMed]

118. Karatan, E.; Watnick, P. Signals, Regulatory Networks, and Materials That Build and Break Bacterial Biofilms. *Microbiol. Mol. Biol. Rev.* **2009**, *73*, 310–347. [CrossRef] [PubMed]

119. Lemire, J.A.; Harrison, J.J.; Turner, R.J. Antimicrobial activity of metals: Mechanisms, molecular targets and applications. *Nat. Rev. Microbiol.* **2013**, *11*, 371–384. [CrossRef] [PubMed]

120. Donlan, R.M. Biofilm elimination on intravascular catheters: Important considerations for the infectious disease practitioner. *Clin. Infect. Dis.* **2011**, *52*, 1038–1045. [CrossRef] [PubMed]

121. Lai, N.M.; Chaiyakunapruk, N.; Lai, N.A.; O'Riordan, E.; Pau, W.S.; Saint, S. Catheter impregnation, coating or bonding for reducing central venous catheter-related infections in adults. *Cochrane Database Syst. Rev.* **2016**, *3*, CD007878. [CrossRef] [PubMed]

122. McGuffie, M.J.; Hong, J.; Bahng, J.H.; Glynos, E.; Green, P.F.; Kotov, N.A.; Younger, J.G.; VanEpps, J.S. Zinc oxide nanoparticle suspensions and layer-by-layer coatings inhibit staphylococcal growth. *Nanomedicine* **2016**, *12*, 33–42. [CrossRef] [PubMed]

123. Song, C.; Yang, C.; Sun, X.; Xia, P.; Qin, J.; Guo, B.; Wang, S. Influences of graphene oxide on biofilm formation of gram-negative and gram-positive bacteria. *Environ. Sci. Pollut. Res.* **2018**, *25*, 2853–2860. [CrossRef] [PubMed]

124. Mao, B.; Cheng, L.; Wang, S.; Zhou, J.; Deng, L. Combat biofilm by bacteriostatic aptamer-functionalized graphene oxide. *Biotechnol. App. Biochem.* **2017**. [CrossRef] [PubMed]

125. Das, M.R.; Sarma, R.K.; Saikia, R.; Kale, V.S.; Shelke, M.V.; Sengupta, P. Synthesis of silver nanoparticles in an aqueous suspension of graphene oxide sheets and its antimicrobial activity. *Colloids Surf. B Biointerfaces* **2011**, *83*, 16–22. [CrossRef] [PubMed]

126. Ruiz, O.N.; Fernando, K.A.S.; Wang, B.; Brown, N.A.; Luo, P.G.; McNamara, P.G.; Vangsness, M.; Sun, J.; Bunker, C.E. Graphene Oxide: A Non Specific Enhancer of Cellular Growth. *ACS Nano* **2011**, *5*, 8100–8107. [CrossRef] [PubMed]

127. Some, S.; Sohn, J.S.; Kim, J.; Lee, S.H.; Lee, S.C.; Lee, J.; Shackery, I.; Kim, S.K.; Kim, S.H.; Choi, N.; et al. Graphene-Iodine Nanocomposites: Highly Potent Bacterial Inhibitors that are Bio-compatible with Human Cells. *Sci. Rep.* **2016**, *6*, 20015. [CrossRef] [PubMed]

128. Wu, P.C.; Chen, H.H.; Chen, S.Y.; Wang, W.L.; Yang, K.L.; Huang, C.H.; Kao, H.F.; Chang, J.C.; Hsu, C.L.L.; Wang, J.Y.; et al. Graphene oxide conjugated with polymers: A study of culture condition to determine whether a bacterial growth stimulant or an antimicrobial agent? *J. Nanobiotechnol.* **2018**, *16*, 1. [CrossRef] [PubMed]

129. Zita Gomes, R.; De Vasconcelos, M.R.; Lopes Guerra, I.M.; De Almeida, R.A.B.; De Campos Felino, A.C. Implant Stability in the Posterior Maxilla: A Controlled Clinical Trial. *Biomed. Res. Int.* **2017**, *2017*, 6825213. [CrossRef] [PubMed]

130. Ren, N.; Li, J.; Qiu, J.; Yan, M.; Liu, H.; Ji, D.; Huang, J.; Yu, J.; Liu, H. Growth and accelerated differentiation of mesenchymal stem cells on graphene-oxide-coated titanate with dexamethasone on surface of titanium implants. *Dent. Mater.* **2017**, *33*, 525–535. [CrossRef] [PubMed]

131. Dhaliwal, J.S.; Marulanda, J.; Li, J.; Alebrahim, S.; Feine, J.S.; Murshed, M. In vitro comparison of two titanium dental implant surface treatments: 3M™ESPE™ MDIs versus Ankylos®. *Int. J. Implant Dent.* **2017**, *3*, 27. [CrossRef] [PubMed]

132. Jung, H.S.; Lee, T.; Kwon, I.K.; Kim, H.S.; Hahn, S.K.; Lee, C.S. Surface Modification of Multipass Caliber-Rolled Ti Alloy with Dexamethasone-Loaded Graphene for Dental Applications. *ACS Appl. Mater. Interfaces* **2015**, *7*, 9598–9607. [CrossRef] [PubMed]

133. Kim, S.E.; Song, S.H.; Yun, Y.P.; Choi, B.J.; Kwon, I.K.; Bae, M.S.; Moon, H.J.; Kwon, Y.D. The effect of immobilization of heparin and bone morphogenic protein-2 (BMP-2) to titanium surfaces on inflammation and osteoblast function. *Biomaterials* **2011**, *32*, 366–373. [CrossRef] [PubMed]

134. Bae, S.E.; Choi, J.; Joung, Y.K.; Park, K.; Han, D.K. Controlled release of bone morphogenetic protein (BMP)-2 from nanocomplex incorporated on hydroxyapatite-formed titanium surface. *J. Controll. Release* **2012**, *160*, 676–684. [CrossRef] [PubMed]

135. Hong, H.S.; Lee, J.; Lee, E.; Kwon, Y.S.; Lee, E.; Ahn, W.; Jiang, M.H.; Kim, J.C.; Son, Y. A new role of substance P as an injury-inducible messenger for mobilization of CD29(+) stromal-like cells. *Nat. Med.* **2009**, *15*, 425–443. [CrossRef] [PubMed]

136. La, W.G.; Jin, M.; Park, S.; Yoon, H.H.; Jeong, G.J.; Bhang, S.H.; Park, H.; Char, K.; Kim, B.S. Delivery of bone morphogenetic protein-2 and substance P using graphene oxide for bone regeneration. *Int. J. Nanomed.* **2014**, *9*, 107–116.

137. Qian, W.; Qiu, J.; Su, J.; Liu, J. Minocycline hydrochloride loaded on titanium by graphene oxide: An excellent antibacterial platform with the synergistic effect of contact-killing and release-killing. *Biomater. Sci.* **2018**, *6*, 304–313. [CrossRef] [PubMed]

138. Jianfeng, J.; Li, Z.; Mengqi, S.; Yumei, Z.; Qintao, W. Ti-GO-Ag nanocomposite: The effect of content level on the antimicrobial activity and cytotoxicity. *Int. J. of Nanomedicine* **2017**, *12*, 4209–4224.

139. Zhou, Q.; Yang, P.; Li, X.; Liu, H.; Ge, S. Bioactivity of periodontal ligament stem cells on sodium titanate coated with graphene oxide. *Sci. Rep.* **2016**, *6*, 19343. [CrossRef] [PubMed]

140. Cucchi, A.; Ghensi, P. Vertical Guided Bone Regeneration using Titanium-reinforced d-PTFE Membrane and Prehydrated Corticocancellous Bone Graft. *Open Dent. J.* **2014**, *8*, 194–200. [CrossRef] [PubMed]

141. Gottlow, J.; Nyman, S.; Karring, T.; Lindhe, J. New attachment formation as the result of controlled tissue regeneration. *J. Clin. Periodontol.* **1984**, *11*, 494–503. [CrossRef] [PubMed]

142. Wang, J.; Wang, L.; Zhou, Z.; Lai, H.; Xu, P.; Liao, L.; Wei, J. Biodegradable Polymer Membranes Applied in Guided Bone/Tissue Regeneration: A Review. *Polymers* **2016**, *8*, 115. [CrossRef]

143. Scantlebury, T.V. 1982–1992: A decade of technology development for guided tissue regeneration. *J. Periodontol.* **1993**, *64*, 1129–1137. [CrossRef] [PubMed]

144. De Marco, P.; Zara, S.; De Colli, M.; Radunovic, M.; Lazović, V.; Ettorre, V.; Di Crescenzo, A.; Piattelli, A.; Cataldi, A.; Fontana, A. Graphene oxide improves the biocompatibility of collagen membranes in an in vitro model of human primary gingival fibroblasts. *Biomed. Mater.* **2017**, *12*, 055005. [CrossRef] [PubMed]

145. Radunovic, M.; De Colli, M.; De Marco, P.; Di Nisio, C.; Fontana, A.; Piattelli, A.; Cataldi, A.; Zara, S. Graphene oxide enrichment of collagen membranes improves DPSCs differentiation and controls inflammation occurrence. *J. Biomed. Mater. Res. A* **2017**, *105*, 2312–2320. [CrossRef] [PubMed]

146. Bregnocchi, A.; Zanni, E.; Uccelletti, D.; Marra, F.; Cavallini, D.; De Angelis, F.; De Bellis, G.; Bossù, M.; Ierardo, G.; Polimeni, A.; Sarto, M.S. Graphene-based dental adhesive with anti-biofilm activity. *J. Nanobiotechnol.* **2017**, *15*, 89. [CrossRef] [PubMed]

147. Zanni, E.; Chandraiahgari, C.R.; De Bellis, G.; Montereali, M.R.; Armiento, G.; Ballirano, P.; Polimeni, A.; Sarto, M.S.; Uccelletti, D. Zinc Oxide Nanorods-Decorated Graphene Nanoplatelets: A Promising Antimicrobial Agent against the Cariogenic Bacterium Streptococcus mutans. *Nanomaterials* **2016**, 6. [CrossRef] [PubMed]

148. Biris, A.R.; Pruneanu, S.; Pogacean, F.; Lazar, M.; Borodi, G.; Ardelean, S.; Dervishi, E.; Watanabe, F.; Biris, A.S. Few-layer graphene sheets with embedded gold nanoparticles for electrochemical analysis of adenine. *Int. J. Nanomedicine.* **2013**, *8*, 1429–1438. [CrossRef] [PubMed]

149. Sarosi, C.; Biris, A.R.; Antoniac, A.; Boboia, S.; Alb, C.; Antoniac, I.; Moldovan, M. The nanofiller effect on properties of experimental graphene dental nanocomposites. *J. Adhes Sci. Technol.* **2016**, *30*, 1779–1794.

150. Li, S.; Zhuanjun, Y.; Youxin, D.; Junyan, Z.; Bin, L. Improvement of the mechanical, tribological andantibacterial properties of glass ionomer cements by fluorinated graphene. *Dent. Mater.* **2018**, *S0109*, 30971–30975.

151. Carey, C.M. Tooth whitening: What we know now. *J. Evid. Based Dent. Pract.* **2014**, *14*, 70–76. [CrossRef] [PubMed]

152. Su, I.H.; Lee, C.F.; Su, Y.P.; Wang, L.H. Evaluating a Cobalt-Tetraphenylporphyrin Complex, Functionalized with a Reduced Graphene Oxide Nanocomposite, for Improved Tooth Whitening. *J. Esthet. Restor. Dent.* **2016**, *5*, 321–329. [CrossRef] [PubMed]

153. Skoda, M.; Dudek, I.; Jarosz, A.; Szukiewicz, D. Graphene: One material, many possibilities- Application difficulties in biological systems. *J. Nanomater.* **2014**. [CrossRef]

154. Rodríguez-Pérez, L.; Herranz, M.A.; Martín, N. The chemistry of pristine graphene. *Chem. Commun.* **2013**, *49*, 3721–3735. [CrossRef] [PubMed]

155. Gao, X.; Wang, Y.; Liu, X.; Chan, L.T.; Irle, S.; Zhao, Y.; Zhang, S.B. Regioselectivity contro of graphene functionalization by ripples. *Phys. Chem. Chem. Phys.* **2011**, *13*, 19449–19453. [CrossRef] [PubMed]

156. Zhang, X.; Yin, J.; Peng, C.; Hu, W.; Zhu, Z.; Li, W.; Fan, C.; Huang, Q. Distribution and biocompatibility studies of graphene oxide in mice after intravenous administration. *Carbon* **2011**, *49*, 986–995. [CrossRef]

nanomaterials

MDPI

Article

Porcine Dental Epithelial Cells Differentiated in a Cell Sheet Constructed by Magnetic Nanotechnology

Wataru Koto, Yoshinori Shinohara *, Kazuyuki Kitamura, Takanori Wachi, Seicho Makihira and Kiyoshi Koyano

Section of Fixed Prosthodontics, Department of Oral Rehabilitation, Faculty of Dental Science, Kyushu University, Fukuoka 812-8582, Japan; k.wataru@dent.kyushu-u.ac.jp (W.K.); k.kitamura@dent.kyushu-u.ac.jp (K.K.); wachi@dent.kyushu-u.ac.jp (T.W.); makihira@dent.kyushu-u.ac.jp (S.M.); koyano@dent.kyushu-u.ac.jp (K.K.)
* Correspondence: sinohara@dent.kyushu-u.ac.jp; Tel.: +81-92-642-6371

Received: 14 September 2017; Accepted: 9 October 2017; Published: 13 October 2017

Abstract: Magnetic nanoparticles (MNPs) are widely used in medical examinations, treatments, and basic research, including magnetic resonance imaging, drug delivery systems, and tissue engineering. In this study, MNPs with magnetic force were applied to tissue engineering for dental enamel regeneration. The internalization of MNPs into the odontogenic cells was observed by transmission electron microscopy. A combined cell sheet consisting of dental epithelial cells (DECs) and dental mesenchymal cells (DMCs) (CC sheet) was constructed using magnetic force-based tissue engineering technology. The result of the iron staining indicated that MNPs were distributed ubiquitously over the CC sheet. mRNA expression of enamel differentiation and basement membrane markers was examined in the CC sheet. Immunostaining showed Collagen IV expression at the border region between DEC and DMC layers in the CC sheet. These results revealed that epithelial–mesenchymal interactions between DEC and DMC layers were caused by bringing DECs close to DMCs mechanically by magnetic force. Our study suggests that the microenvironment in the CC sheet might be similar to that during the developmental stage of a tooth bud. In conclusion, a CC sheet employing MNPs could be developed as a novel and unique graft for artificially regenerating dental enamel.

Keywords: magnetic nanoparticles; nanotechnology; cell sheet; odontogenic cells; epithelial-mesenchymal interactions; dental enamel regeneration

---

1. Introduction

Magnetic nanoparticles (MNPs) are widely used in medical treatments and clinical and basic research because of their unique features, such as superparamagnetic properties, biocompatibility, and non-toxicity. MNPs are used as contrast media and have been extensively used in the field of magnetic resonance imaging due to their better imaging resolution and sensitivity [1,2]. Since the technology of coating molecules and antibodies on the surface of MNPs was well established, MNPs have been applied in drug delivery systems and hyperthermia in cancer therapy [3–8]. MNPs can also be used to decorate biological nanofibers [9]. These nanofibers decorated with MNPs can be induced to form biological nanofiber assemblies on the centimeter scale by applying magnetic force. These long-range-ordered assemblies can be used as scaffolds in tissue engineering to encourage the adhesion, proliferation, and differentiation of various cells [9].

In the field of regenerative medicine, cell sheets created by magnetic force-based tissue engineering technology (Mag-TE) have been applied to the regeneration of many organs, including skin, bone, liver, and heart. Such cell sheets prepared by Mag-TE are reported to be functional and have favorable structures [10–14]. It is reported that multilayered keratinocyte sheets can be constructed by using Mag-TE system and be harvested without enzymatic treatment [10]. The transplantation of mesenchymal stem cell (MSC) sheets constructed by the Mag-TE system induced bone formation in bone defect areas in the crania of nude rats, and it is suggested that these MSC sheets are useful for bone tissue engineering [11]. By using the Mag-TE system, human aortic endothelial cells (HAECs) adhered to form a layered construct with tight and close contact on rat hepatocyte monolayers [12]. In this co-cultured construct, albumin secretion by hepatocytes was significantly enhanced compared with that in homotypic cultures of hepatocytes or heterotypic co-cultures of hepatocytes and HAECs without using the Mag-TE system. In heart revision, the presence of gap junctions and electrical connections were found within the cardiomyocytes sheets using the Mag-TE system [13]. In addition, it was demonstrated that human iPS cell-derived fetal liver kinase-1 positive cell sheets accelerated revascularization of ischemic hindlimbs in nude mice [14]. These several reports suggest that MNPs and the Mag-TE system are useful for regenerative medicine.

There has been great interest in the regeneration of tooth and periodontal tissue. Although attempts at reconstruction of dentin, periodontal ligaments, and alveolar bone have been relatively successful, regeneration of dental enamel (DE) remains difficult because of problems due to its specific developmental process [15,16]. It has been reported that DE is formed and matured by epithelial–mesenchymal interactions. Similar to the development of tooth germ, the establishment of such constant epithelial–mesenchymal interactions depends heavily on the positional relationship between dental epithelial cells (DECs) and dental mesenchymal cells (DMCs) [15,17]. However, no method has been developed to address the issue of cell position. Therefore, we attempted to solve this problem by focusing on MNPs, materials that can be used to control the arrangement of DECs and DMCs [18], and hypothesized that we could imitate the microenvironment in the developmental stage of a tooth bud in vitro.

In this study, we investigated the effects of MNPs and exposure to magnetic force on DECs and DMCs obtained from porcine tooth germ. We constructed a combined cell sheet consisting of DECs and DMCs (CC sheet) using the Mag-TE system, and examined the expression of differentiation and basement membrane markers in this cell sheet.

2. Results

2.1. Internalization of MNPs in Odontogenic Cells

Internalization of MNPs in odontogenic cells was evident from transmission electron microscopy (TEM) microphotographs (Figure 1). The presence of MNPs in the cytoplasm was confirmed. The diameters of MNPs were also measured on TEM images at higher magnification and the average was approximately 20 nm (data not shown).

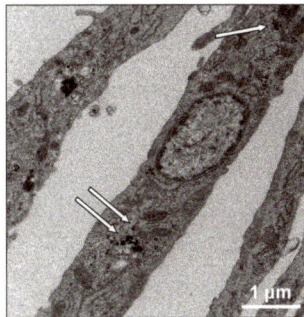

Figure 1. Transmission electron microscopy (TEM) image of odontogenic cells treated with 100 pg-magnetite/cell of magnetic nanoparticles (MNPs) ($\times$5000). White arrows indicate intracellular MNPs. Scale bar = 1 μm.

2.2. Cytotoxicity of MNPs in DECs and DMCs

The cytotoxic effects of MNPs on DECs and DMCs were assessed by MTS assays. MNPs at 0, 50, 100, 150, or 300 pg-magnetite/cell were added to DECs and DMCs at confluency. There were no significant changes in the absorbances at all concentrations for both DECs and DMCs ($p > 0.05$, one-way analysis of variance (ANOVA) with Tukey's multiple range test) (Figure 2).

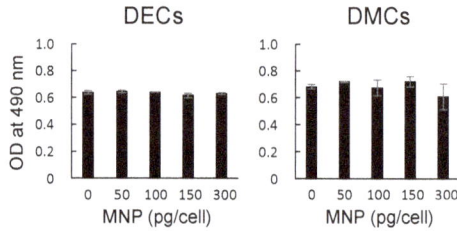

Figure 2. Cytotoxic effects of MNPs on dental epithelial cells (DECs) and dental mesenchymal cells (DMCs) assessed by MTS assays. MNPs at 0, 50, 100, 150, or 300 pg-magnetite/cell were added to the cells at confluency. After the cells were maintained for 24 h in the presence or absence of MNPs, the MTS assay was performed. $p > 0.05$, ANOVA.

2.3. Effects of MNPs and/or Magnetic Force on mRNA Expression of Enamel Matrix Genes in DECs

The mRNA expression profiles of *Amelogenin* (*AMEL*), *Enamelin* (*ENAM*), and *Ameloblastin* (*AMBN*) in DECs were investigated after the cells were exposed to MNPs and/or magnetic force for 12 and 24 h. Depending on the presence of MNPs or magnetic force, experimental groups were classified into four subgroups as shown in Table 1. There were no significant changes in the levels of all mRNAs in groups 1–4 when the magnetic force was applied for 12 h ($p > 0.05$, ANOVA). Group 4 with 24 h of exposure to magnetic force showed significant increases in the mRNA expression of *AMBN* compared with groups 1–3 (* $p < 0.05$, ANOVA), although there were no significant changes in the levels of *AMEL* or *ENAM* mRNAs ($p > 0.05$, ANOVA) (Figure 3).

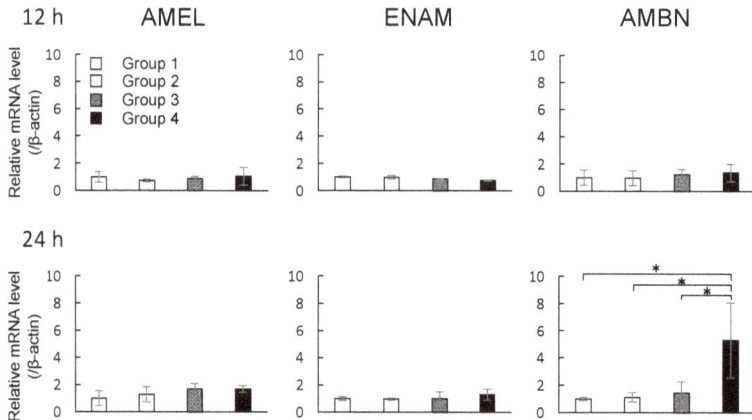

Figure 3. Effects of MNPs and/or magnetic force on the expression levels of *AMEL*, *ENAM*, and *AMBN* in cultured DECs examined by real-time reverse transcriptase polymerase chain reaction (RT-PCR). Real-time RT-PCR data were normalized to the expression levels of *β-actin* mRNA. Independent experiments were repeated twice. Data represent the mean ± standard deviation (SD) of triplicate samples. * $p < 0.05$, ANOVA.

Table 1. Classification of experimental groups depending on the presence of MNPs or magnetic force loading.

Classification of The Experimental Groups	Group 1	Group 2	Group 3	Group 4
MNPs	-	-	+	+
Magnetic Force	-	+	-	+

2.4. Effects of MNPs and/or Magnetic Force on the mRNA Expression of Dentin-Related Genes in DMCs

The mRNA expression profiles of *Runt-related transcription factor 2* (*RUNX2*), *CollagenIα2* (*COL1α2*), and *Dentin sialophosphoprotein* (*DSPP*) in DMCs were investigated after the cells were exposed to MNPs and/or magnetic force for 12 and 24 h. Depending on the presence of MNPs or magnetic force, experimental groups were classified into four subgroups as shown Table 1. There were no significant changes in the levels of *RUNX2*, *COL1α2*, or *DSPP* mRNAs in groups 1–4 when the magnetic force was applied for 12 or 24 h ($p > 0.05$, ANOVA) (Figure 4).

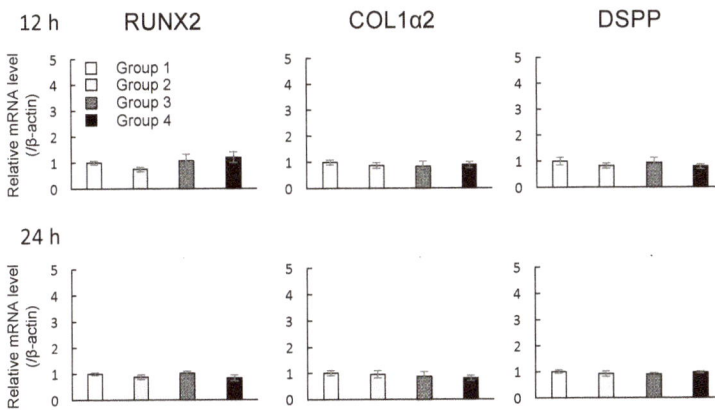

Figure 4. Effects of MNPs and/or magnetic force on the expression levels of *RUNX2*, *COL1α2*, and *DSPP* in cultured DMCs examined by real-time RT-PCR. Real-time RT-PCR data were normalized to the expression levels of *β-actin* mRNA. Independent experiments were repeated twice. Data represent the mean ± SD of triplicate samples. $p > 0.05$, ANOVA.

2.5. Construction of DEC and DMC Sheets Using the Mag-TE System

We attempted to construct DEC and DMC sheets using the Mag-TE system [14]. By applying magnetic force, both DECs and DMCs labeled with MNPs migrated to form a sheet-like structure in the middle of the culture well after 24 h of incubation (Figure 5A). The cell sheets were circular with diameters of 8 mm and appeared to have the same structure in repeated experiments. These sheets were embedded in paraffin, sectioned, and observed after hematoxylin and eosin (HE) staining. The cell sheets had a multilayered structure of approximately 20 μm in thickness (Figure 5B).

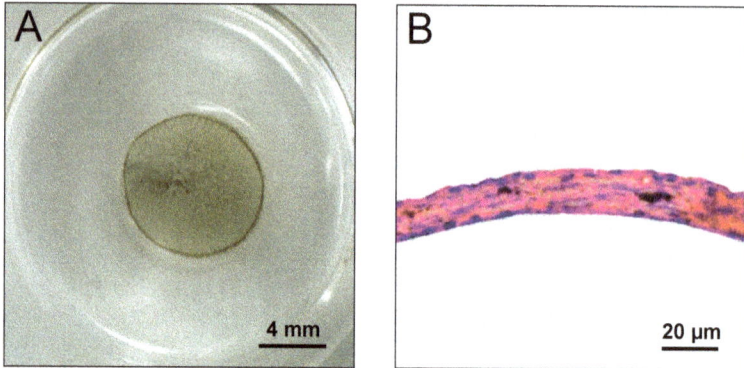

Figure 5. Photographs of a DEC sheet created by the Mag-TE system. DECs labeled with MNPs were seeded into a 24-well ultra low attachment culture plate (Corning Inc., Corning, NY, USA) at 2×10^6 cells/plate. A cylindrical neodymium magnet (magnetic force, 4110 G) was placed on the reverse side of the ultra low attachment plate, and the cells were cultured for 24 h. After the culture, the cell sheet was constructed (**A**). Microscopic image of an HE-stained cross section of a DEC sheet (×100) (**B**).

2.6. Localization of MNPs in CC Sheet

A CC sheet was constructed using the Mag-TE system. The CC sheet had a multilayered structure approximately 40 µm in thickness. Iron staining was performed to visualize the localization of MNPs in CC sheet. MNPs were distributed ubiquitously over the CC sheet (Figure 6).

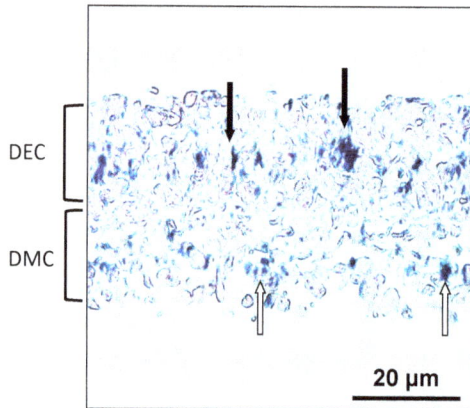

Figure 6. Microscopic observation of a CC sheet after iron staining (×200). Filled arrows indicate MNPs in the DEC layer. White arrows indicate MNPs in the DMC layer. Scale bar = 20 µm.

2.7. Expression of mRNAs Encoding Enamel Matrix- and Dentin-Related Genes in CC Sheet

The mRNA expression levels of *AMEL*, *ENAM*, *AMBN*, *RUNX2*, and *Collagen IV 1* (*COL4 1*) in CC sheet were higher than those in the control (* $p < 0.05$, ** $p < 0.01$), although there were no significant changes in the levels of *COL1α2* or *DSPP* mRNA expression ($p > 0.05$) (Figure 7).

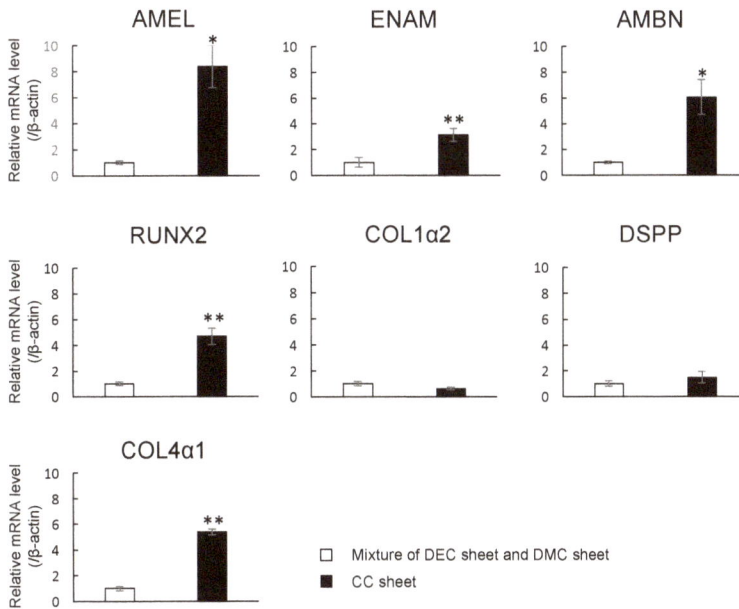

Figure 7. mRNA expression levels of *AMEL, ENAM, AMBN, RUNX2, COL1α2, DSPP,* and *COL4α1* in CC sheet examined by real-time RT-PCR. A mixture of DEC and DMC sheets prepared separately were used as a control. Real-time RT-PCR data were normalized to the expression levels of *β-actin* mRNA. Independent experiments were repeated twice. Data represent the mean ± SD of triplicate samples. * $p < 0.05$. ** $p < 0.01$.

2.8. Localization of Collagen IV in CC Sheet

Immunofluorescence staining was performed to assess the localization of Collagen IV(COL4) that is expressed in the basement membrane of presecretory and late mature stage tooth germs. The nuclei of live cells were stained with 4′,6-diamidino-2-phenylindole (DAPI) (Figure 8A). COL4 expression was localized markedly in the middle region of the CC sheet (Figure 8B,C).

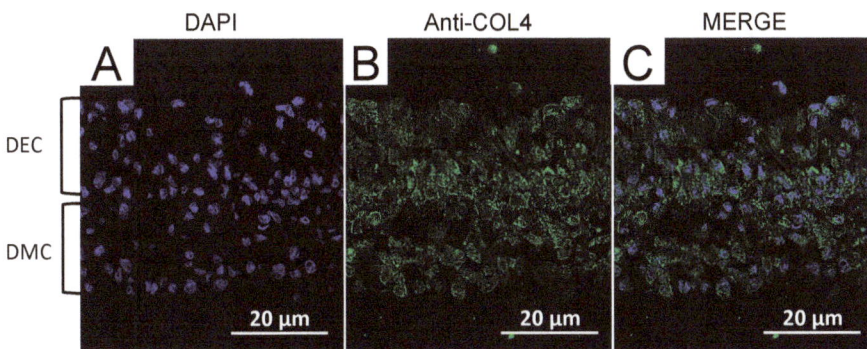

Figure 8. Fluorescence microscopic observation of a CC sheet (×200). DAPI (blue) was used to stain nuclei (**A**). COL4 was stained with a specific antibody. COL4-expressing cells (green) were identified around the border of DEC and DMC layers (**B,C**). Scale bars = 20 µm.

3. Discussion

Recently, MNPs have been attracting a lot of attention in medical field. Due to their special and unique properties, they have been studied and applied clinically in various medical areas, including regeneration. In the field of regeneration, tooth regeneration using tissue engineering techniques has been studied and developed actively, but the techniques or methods are not well established [19]. Tooth development is regulated by intricate interactions between DECs and DMCs, but it has not been completely elucidated yet [15,17]. In particular, DE is formed through a complicated development process [20]. Therefore, many problems remain in the field of enamel regeneration. For example, although a collagen sponge has been used as a scaffold to transplant a mixture of DECs and DMCs, the success rate of tooth production is not very high [21]. This result indicates that it is essential for enamel regeneration to control the three-dimensional position of DECs and DMCs to be similar to the tooth developmental process. Although DECs and DMCs are aligned and in contact via the basement membrane throughout tooth development [22], it is difficult to mimic this microenvironment in vitro. Therefore, we focused on a cell sheet and determined whether we could reproduce the cell arrangement and three-dimensional cell position in vitro by layering two cell sheets of two different cell types. MNPs have an ability to make different cells cohere together by applying magnetic force [12]. This is the reason why we adopted the method employing MNPs among the various methods used to prepare cell sheets, such as using temperature-responsive cell culture dishes [23,24] and collagen gels [25].

Although MNPs are applied to regenerative medicine of many organs and are known as a biocompatible material [10–14,26], neither the influence of MNPs on odontogenic cells nor their applicability to regenerative dentistry have been reported. First, we examined the internalization of MNPs in the odontogenic cells by TEM. The TEM imaging revealed that MNPs were localized in the cytoplasm, and it is suggested that MNPs were taken up in the odontogenic cells (Figure 1). This result was similar to previous research which reported the uptake of MNPs in different cell types [27–29]. The result obtained in the present study and the previous reports [27–29] suggest that MNPs can be used as carriers to deliver biologically active agents, such as growth factors and nanovectors, to odontogenic cells for control of histodifferentiation and organ generation in vivo [30,31]. We further investigated the cytotoxicity of MNPs in DECs and DMCs. The results of MTS assays suggested that MNPs had no cytotoxicity in DECs or DMCs up to 300 pg-magnetite/cell (Figure 2). Therefore, we believed that we could apply MNPs to DE regeneration. In this study, a DEC sheet, DMC sheet, and combined cell sheet of DECs and DMCs (CC sheet) were constructed using the Mag-TE system. The CC sheet was constructed by directly laminating magnetically labeled DECs on a prepared DMC sheet by magnetic force loading. Although cell sheets can be piled up by a method involving temperature-responsive cell culture dishes, the cell sheets prepared in this way must be peeled off and placed on another cell sheet. As cell sheets are thin, they require skill in handling to multi-layer sheets one by one [23,24,32,33]. Therefore, the multi-layering technique with the Mag-TE system will be useful because the manipulation required to remove a cell sheet can be avoided. Furthermore, in the method with temperature-responsive cell culture dishes, it is difficult to cohere two cell sheets together [34], whereas in our method with the Mag-TE system, layers of DECs and DMCs could be brought into close proximity by magnetic force. It is suggested that epithelial–mesenchymal interaction between the DEC layer and DMC layer is more likely to occur when the Mag-TE system is applied. In this CC sheet, the mRNA levels of *AMEL, ENAM, AMBN, RUNX2,* and *COL4α1* were significantly higher than those in a mixture of DEC and DMC sheets prepared separately. *AMEL, ENAM,* and *AMBN* are differentiation markers of DECs which are known to differentiate into ameloblasts [35–41]. Therefore, our results suggest that DECs differentiated into enamel-secreting ameloblasts [42], and imply that epithelial–mesenchymal interactions proceeded by bringing DEC and DMC layers physically close to each other by magnetic force. *RUNX2* is a differentiation marker of osteoblasts and odontoblasts [43]. The increase in *RUNX2* mRNA in the CC sheet suggests that DMCs differentiated into odontoblasts because of the enhancement of epithelial–mesenchymal interactions. The result of the iron staining (Figure 6) indicated that the MNPs were distributed ubiquitously over the CC sheet. This suggests that the magnetic force can be loaded uniformly to the CC sheet, and that epithelial–mesenchymal interactions may occur homogeneously in the CC sheet. The basement membrane, which interfaces with the dental epithelium and papilla mesenchyme [44,45], is involved in epithelial–mesenchymal interactions [46,47]. COL4 is a membrane form of collagen expressed in tooth bud basement membrane at the presecretory stage and reported to be related to ameloblast differentiation and tooth development [39,45,48,49]. These studies suggest that the interactions mediated through the basement membrane and the basement membrane components such as COL4 are essential for DEC growth and differentiation to form the proper shape and size of the tooth. Taken together with these previous findings, the results in the present study, which showed enhancement of *COL4α1* mRNA expression and localization of COL4 by immunofluorescence staining in CC sheet, suggest the existence of a basement membrane between the DEC and DMC layers (Figure 8B,C).

Our results collectively support the hypothesis that epithelial–mesenchymal interactions between DEC and DMC layers were induced by the Mag-TE system. Further study is required to verify the utility of CC sheets employing MNPs as a cutting edge technology, including tissue regenerating experiments in vivo.

4. Materials and Methods

4.1. Materials

MNPs (Nano3D Biosciences, Houston, TX, USA) and a neodymium magnet (diameter = 25 mm, magnetic force = 4110 G; NeoMag Co., Ltd., Chiba, Japan) were used in this study. MNPs (0.05%, *w*/*v*) coated with poly-L-lysine were diluted to 0.001% (*w*/*v*) with Dulbecco's modified Eagle's medium (DMEM; Nacalai Tesque Inc., Kyoto, Japan) containing 10% fetal bovine serum (FBS; Hyclone®; Thermo Fisher Scientific Inc., Waltham, MA, USA) and 1% antibiotics (Gibco® Antibiotic-Antimycotic; Thermo Fisher Scientific Inc., Waltham, MA, USA).

4.2. Primary Culture of Odontogenic Cells In Vitro

DECs and DMCs were isolated from third molar tooth buds of 6-month-old porcine lower jaws (Fukuokashokunikuhanbai Co. Ltd., Fukuoka, Japan) and cultured as described previously [16]. Single cell suspensions of DECs and DMCs were seeded in 10-cm dishes (Corning Inc., Corning, NY, USA) and maintained in DMEM containing 10% FBS and 1% antibiotics at 37 °C with 5% CO_2.

4.3. Transmission Electron Microscopy (TEM)

Odontogenic cells were incubated with MNPs (100 pg-magnetite/cell) for 24 h. After incubation, cells were collected, washed three times with phosphate buffered saline (PBS), and fixed with a fix buffer (2.5% glutaraldehyde, 0.1 M sucrose, 3 mM $CaCl_2$, and 0.1 M sodium cacodylate, pH 7.4) for 2 h. Next, the cells were rinsed in 0.1 M sodium phosphate for 15 min at room temperature. The cells were then postfixed with 1% OsO_4 for 1.5 h and rinsed in 0.1 M sodium phosphate overnight at 4 °C. Subsequently, the cells were dehydrated in graded alcohol concentrations and propylene oxide, embedded in epoxy resin, and incubated for 2 days at 65 °C. Ultrathin sections (80 nm) were prepared by a Leica EM UC7 (Leica microsystems GmbH, Wetzlar, Germany). Finally, the sections were stained with 2% uranyl acetate and lead acetate for 5 and 10 min, respectively. The sections were observed using a Tecnai-20 (FEI Co., Hillsboro, OR, USA).

4.4. Cytotoxicity Assay

The cytotoxicity of MNPs in DECs and DMCs was assessed by MTS assays. In brief, DECs or DMCs were seeded onto a 96-well flat bottom cell culture plate (Corning Inc., Corning, NY, USA) at 3×10^3 cells/well and incubated with MNPs at 0, 50, 100, 150, or 300 pg-magnetite/cell. After 24 h of incubation at 37 °C, the cells were exposed to an MTS solution for 24 h. Suspended cells were removed by gentle rinsing with phosphate-buffered saline (PBS), and the number of adherent cells remaining in each well was quantified using a coupled enzymatic assay resulting in conversion of a tetrazolium salt into a red formazan product (CellTiter 96 Aqueous Non-Radioactive Cell Proliferation Assay, Promega, Madison, WI, USA). Recording the absorbance at 490 nm in the MTS assay was carried out using a microplate reader (infinite M200, Tecan Japan Co., Ltd., Kanagawa, Japan).

4.5. Construction of Cell Sheets

DEC and DMC sheets were constructed using the Mag-TE system [10,14,26]. To magnetically label the cells, DECs and DMCs were first incubated with MNPs at 100 pg-magnetite/cell. After 6 h of incubation at 37 °C, both cell types labeled with MNPs were seeded into a 24-well ultra low attachment cell culture plate (Corning Inc., Corning, NY, USA) at 2×10^6 cells/plate. The cylindrical neodymium magnet (4110 G) was placed on the reverse side of the ultra low attachment plate, and the cells were cultured for an additional 24 h. The external magnetic field was provided perpendicular to the cell layers. Then, the neodymium magnet was removed from the culture plate. A CC sheet was also constructed using the Mag-TE system. First, 2×10^6 magnetically labeled DMCs were seeded into a 24-well ultra low attachment cell culture plate with the external magnetic field provided perpendicular to the cell layers and cultured for 24 h. After generation of the DMC sheet, 2×10^6 magnetically labeled DECs were seeded into the same 24-well ultra low attachment cell culture plate and incubated for another 24 h. Then, the neodymium magnet was removed from the culture plate.

4.6. Real-Time Reverse Transcriptase Polymerase Chain Reaction (RT-PCR) Analyses

Total RNA was extracted using TRIzol reagent (Invitrogen, Carlsbad, CA, USA). First-strand cDNA was synthesized from 100 ng total RNA using ReverTra Ace (Toyobo, Osaka, Japan). The cDNA was then amplified by SYBR green 1 DNA polymerase (TAKARA BIO Inc., Shiga, Japan). Real-time RT-PCR analyses of *Amelogenin (AMEL)* (Sigma-Aldrich Co. LLC., Tokyo, Japan), *Enamelin (ENAM)* (Sigma-Aldrich Co. LLC., Tokyo, Japan), *Ameloblastin (AMBN)* (Sigma-Aldrich Co. LLC., Tokyo, Japan), *Runt-related transcription factor 2 (RUNX2)* (Sigma-Aldrich Co. LLC., Tokyo, Japan), *CollagenIα2 (COL1α2)* (Sigma-Aldrich Co. LLC., Tokyo, Japan), *Dentin sialophosphoprotein (DSPP)* (Sigma-Aldrich Co. LLC., Tokyo, Japan), *CollagenIVα1 (COL4α1)* (BEX CO., LTD., Tokyo, Japan), and *β-actin* (Sigma-Aldrich Co. LLC., Tokyo, Japan) were performed using a Rotor-Gene 6000 (Qiagen, Tokyo, Japan). *β-Actin* was chosen as an internal control to standardize the variability in amplification owing to slight differences in total starting RNA concentrations. Primer and probe sequences are listed in Supplementary Table S1.

4.7. Prussian Blue Staining

To visualize the localization of MNPs in CC sheet, an iron staining kit (Muto pure chemicals Co., Ltd., Tokyo, Japan) was used. A CC sheet constructed by the Mag-TE system was fixed with 4% paraformaldehyde and then rinsed with PBS three times. After dehydrating in graded alcohol concentrations, the samples were embedded in paraffin, and 3 μm-thick serial sections were prepared and mounted on glass slides. Deparaffinized sections were stained with the mixture of filtrated 2% potassium ferrocyanide II/2% hydrochloric acid (ratio 1:1) for 60 min at 37 °C and washed in distilled water. Finally, the sections were counter-stained with 1% safranin O for 5 min, washed three times in distilled water, and air dried.

4.8. Immunohistochemical Observations

A CC sheet constructed by the Mag-TE system was fixed with 4% paraformaldehyde and then rinsed with PBS three times. After dehydrating in graded alcohol concentrations, the samples were embedded in paraffin, and 3 μm-thick serial sections were prepared and mounted on glass slides. Deparaffinized sections were incubated in an antigen retrieval solution (HistoVT One, Nacalai Tesque,

Kyoto, Japan) for 20 min at 90 °C, followed by blocking with EzBlock BSA (ATTO, Tokyo, Japan) for 30 min. The sections were incubated overnight with a primary antibody against COL4 (anti-collagen IV antibody, ab6586, Abcam Co., Ltd., Tokyo, Japan), which is a basement membrane marker, followed by 30 min of incubation with *Alexa Fluor 488*-conjugated secondary donkey anti-rabbit IgG (Thermo Fisher Scientific Inc., Waltham, MA, USA). Then, the sections were gently rinsed three times with PBS and counterstained with 4′,6-diamidino-2-phenylindole (DAPI) (Dojindo, Kumamoto, Japan) for 10 min. Fluorescence images were acquired with a BZ-9000 (Keyence, Osaka, Japan).

4.9. Statistical Analyses

Differences between group averages were assessed by Student's *t*-tests or one-way analysis of variance (ANOVA) with Tukey's multiple range test. SPSS version 20.0.0 (IBM SPSS Statistics, IBM, Tokyo, Japan) was used for statistical analyses.

5. Conclusions

In the present study, MNPs were shown to be taken up in odontogenic cells, and were found to be a biocompatible material for odontogenic cells at a limited concentration. The Mag-TE system demonstrated that the epithelial–mesenchymal interactions and the differentiation stage of DECs could be controlled, which is thought to be important to regenerate DE. Regeneration of DE can be expected in vivo by reconstructing the microenvironment and differentiation stage of odontogenic cells in vitro. Consequently, MNPs may be a promising and unique material for regenerative dentistry.

Supplementary Materials: The following are available online at www.mdpi.com/2079-4991/7/10/322/s1, Table S1: Primers/probes for real-time RT-PCR.

Acknowledgments: This study was supported in part by Grants-in-Aid for Scientific Research from the Japan Society for the Promotion of Science to Yoshinori Shinohara (Grant No. 16K11602) and Seicho Makihira (Grant No. 15K11162). We appreciate the technical support from the Research Support Center, Graduate School of Medical Sciences, Kyushu University. We also appreciate the technical support from the Laboratory for technical support, Medical Institute of Bioregulation, Kyushu University. We also thank Alison Sherwin, PhD, from Edanz Group (www.edanzediting.com) for editing a draft of this manuscript.

Author Contributions: Wataru Koto, Yoshinori Shinohara, and Kiyoshi Koyano conceived and designed the experiments; Wataru Koto performed the experiments; Wataru Koto, Seicho Makihira, Takanori Wachi, and Kazuyuki Kitamura analyzed the data; Wataru Koto wrote the manuscript.

Conflicts of Interest: The authors declare no conflict of interest.

References

1. Gillis, P.; Koenig, S.H. Transverse relaxation of solvent protons induced by magnetized spheres: Application to ferritin, erythrocytes, and magnetite. *Magn. Reson. Med.* **1987**, *5*, 323–345. [CrossRef] [PubMed]
2. Lee, J.H.; Kim, J.W.; Cheon, J. Magnetic nanoparticles for multi-imaging and drug delivery. *Mol. Cells* **2013**, *35*, 274–284. [CrossRef] [PubMed]
3. Suzuki, M.; Honda, H.; Kobayashi, T.; Wakabayashi, T.; Yoshida, J.; Takahashi, M. Development of a target-directed magnetic resonance contrast agent using monoclonal antibody-conjugated magnetic particles. *Noshuyo Byori* **1996**, *13*, 127–132. [PubMed]
4. Mou, X.; Ali, Z.; Li, S.; He, N. Applications of Magnetic Nanoparticles in Targeted Drug Delivery System. *J. Nanosci. Nanotechnol.* **2015**, *15*, 54–62. [CrossRef] [PubMed]
5. Shinkai, M.; Yanase, M.; Honda, H.; Wakabayashi, T.; Yoshida, J.; Kobayashi, T. Intracellular hyperthermia for cancer using magnetite cationic liposomes: In vitro study. *Jpn. J. Cancer Res. Gann* **1996**, *87*, 1179–1183. [CrossRef] [PubMed]
6. Yanase, M.; Shinkai, M.; Honda, H.; Wakabayashi, T.; Yoshida, J.; Kobayashi, T. Intracellular hyperthermia for cancer using magnetite cationic liposomes: An in vivo study. *Jpn. J. Cancer Res. Gann* **1998**, *89*, 463–469. [CrossRef] [PubMed]

7. Suzuki, M.; Shinkai, M.; Honda, H.; Kobayashi, T. Anticancer effect and immune induction by hyperthermia of malignant melanoma using magnetite cationic liposomes. *Melanoma Res.* **2003**, *13*, 129–135. [CrossRef] [PubMed]

8. Obaidat, I.M.; Issa, B.; Haik, Y. Magnetic Properties of Magnetic Nanoparticles for Efficient Hyperthermia. *Nanomaterials* **2015**, *5*, 63–89. [CrossRef] [PubMed]

9. Cao, B.; Zhu, Y.; Wang, L.; Mao, C. Controlled alignment of filamentous supramolecular assemblies of biomolecules into centimeter-scale highly ordered patterns by using nature-inspired magnetic guidance. *Angew. Chem. Int. Ed. Engl.* **2013**, *52*, 11750–11754. [CrossRef] [PubMed]

10. Ito, A.; Hayashida, M.; Honda, H.; Hata, K.; Kagami, H.; Ueda, M.; Kobayashi, T. Construction and harvest of multilayered keratinocyte sheets using magnetite nanoparticles and magnetic force. *Tissue Eng.* **2004**, *10*, 873–880. [CrossRef] [PubMed]

11. Shimizu, K.; Ito, A.; Yoshida, T.; Yamada, Y.; Ueda, M.; Honda, H. Bone tissue engineering with human mesenchymal stem cell sheets constructed using magnetite nanoparticles and magnetic force. *J. Biomed. Mater. Res. Part B Appl. Biomater.* **2007**, *82*, 471–480. [CrossRef] [PubMed]

12. Ito, A.; Takizawa, Y.; Honda, H.; Hata, K.; Kagami, H.; Ueda, M.; Kobayashi, T. Tissue engineering using magnetite nanoparticles and magnetic force: Heterotypic layers of cocultured hepatocytes and endothelial cells. *Tissue Eng.* **2004**, *10*, 833–840. [CrossRef] [PubMed]

13. Shimizu, K.; Ito, A.; Lee, J.K.; Yoshida, T.; Miwa, K.; Ishiguro, H.; Numaguchi, Y.; Murohara, T.; Kodama, I.; Honda, H. Construction of multi-layered cardiomyocyte sheets using magnetite nanoparticles and magnetic force. *Biotechnol. Bioeng.* **2007**, *96*, 803–809. [CrossRef] [PubMed]

14. Kito, T.; Shibata, R.; Ishii, M.; Suzuki, H.; Himeno, T.; Kataoka, Y.; Yamamura, Y.; Yamamoto, T.; Nishio, N.; Ito, S.; et al. iPS cell sheets created by a novel magnetite tissue engineering method for reparative angiogenesis. *Sci. Rep.* **2013**, *3*, 1418. [CrossRef] [PubMed]

15. Thesleff, I. Epithelial-mesenchymal signalling regulating tooth morphogenesis. *J. Cell Sci.* **2003**, *116*, 1647–1648. [CrossRef] [PubMed]

16. Young, C.S.; Terada, S.; Vacanti, J.P.; Honda, M.; Bartlett, J.D.; Yelick, P.C. Tissue engineering of complex tooth structures on biodegradable polymer scaffolds. *J. Dent. Res.* **2002**, *81*, 695–700. [CrossRef] [PubMed]

17. Puthiyaveetil, J.S.; Kota, K.; Chakkarayan, R.; Chakkarayan, J.; Thodiyil, A.K. Epithelial—Mesenchymal Interactions in Tooth Development and the Significant Role of Growth Factors and Genes with Emphasis on Mesenchyme—A Review. *J. Clin. Diagn. Res. JCDR* **2016**, *10*, ZE05–ZE09. [CrossRef] [PubMed]

18. Ito, A.; Akiyama, H.; Kawabe, Y.; Kamihira, M. Magnetic force-based cell patterning using Arg-Gly-Asp (RGD) peptide-conjugated magnetite cationic liposomes. *J. Biosci. Bioeng.* **2007**, *104*, 288–293. [CrossRef] [PubMed]

19. Oshima, M.; Tsuji, T. Functional tooth regenerative therapy: Tooth tissue regeneration and whole-tooth replacement. *Odontology* **2014**, *102*, 123–136. [CrossRef] [PubMed]

20. Simmer, J.P.; Papagerakis, P.; Smith, C.E.; Fisher, D.C.; Rountrey, A.N.; Zheng, L.; Hu, J.C. Regulation of dental enamel shape and hardness. *J. Dent. Res.* **2010**, *89*, 1024–1038. [CrossRef] [PubMed]

21. Sumita, Y.; Honda, M.J.; Ohara, T.; Tsuchiya, S.; Sagara, H.; Kagami, H.; Ueda, M. Performance of collagen sponge as a 3-D scaffold for tooth-tissue engineering. *Biomaterials* **2006**, *27*, 3238–3248. [CrossRef] [PubMed]

22. Reith, E.J.; Boyde, A. The arrangement of ameloblasts on the surface of maturing enamel of the rat incisor tooth. *J. Anat.* **1981**, *133*, 381–388. [PubMed]

23. Kushida, A.; Yamato, M.; Konno, C.; Kikuchi, A.; Sakurai, Y.; Okano, T. Temperature-responsive culture dishes allow nonenzymatic harvest of differentiated Madin-Darby canine kidney (MDCK) cell sheets. *J. Biomed. Mater. Res.* **2000**, *51*, 216–223. [CrossRef]

24. Hirose, M.; Kwon, O.H.; Yamato, M.; Kikuchi, A.; Okano, T. Creation of designed shape cell sheets that are noninvasively harvested and moved onto another surface. *Biomacromolecules* **2000**, *1*, 377–381. [CrossRef] [PubMed]

25. Goto, E.; Mukozawa, M.; Mori, H.; Hara, M. A rolled sheet of collagen gel with cultured Schwann cells: Model of nerve conduit to enhance neurite growth. *J. Biosci. Bioeng.* **2010**, *109*, 512–518. [CrossRef] [PubMed]

26. Ishii, M.; Shibata, R.; Numaguchi, Y.; Kito, T.; Suzuki, H.; Shimizu, K.; Ito, A.; Honda, H.; Murohara, T. Enhanced angiogenesis by transplantation of mesenchymal stem cell sheet created by a novel magnetic tissue engineering method. *Arterioscler. Thromb. Vasc. Biol.* **2011**, *31*, 2210–2215. [CrossRef] [PubMed]

27. Babic, M.; Horak, D.; Trchova, M.; Jendelova, P.; Glogarova, K.; Lesny, P.; Herynek, V.; Hajek, M.; Sykova, E. Poly(L-lysine)-modified iron oxide nanoparticles for stem cell labeling. *Bioconjug. Chem.* **2008**, *19*, 740–750. [CrossRef] [PubMed]

28. Osman, O.; Zanini, L.F.; Frenea-Robin, M.; Dumas-Bouchiat, F.; Dempsey, N.M.; Reyne, G.; Buret, F.; Haddour, N. Monitoring the endocytosis of magnetic nanoparticles by cells using permanent micro-flux sources. *Biomed. Microdevices* **2012**, *14*, 947–954. [CrossRef] [PubMed]

29. Sato, A.; Itcho, N.; Ishiguro, H.; Okamoto, D.; Kobayashi, N.; Kawai, K.; Kasai, H.; Kurioka, D.; Uemura, H.; Kubota, Y.; et al. Magnetic nanoparticles of Fe_3O_4 enhance docetaxel-induced prostate cancer cell death. *Int. J. Nanomed.* **2013**, *8*, 3151–3160. [CrossRef]

30. Cao, B.; Yang, M.; Zhu, Y.; Qu, X.; Mao, C. Stem cells loaded with nanoparticles as a drug carrier for in vivo breast cancer therapy. *Adv. Mater.* **2014**, *26*, 4627–4631. [CrossRef] [PubMed]

31. Magro, M.; Martinello, T.; Bonaiuto, E.; Gomiero, C.; Baratella, D.; Zoppellaro, G.; Cozza, G.; Patruno, M.; Zboril, R.; Vianello, F. Covalently bound DNA on naked iron oxide nanoparticles: Intelligent colloidal nano-vector for cell transfection. *Biochim. Biophys. Acta* **2017**. [CrossRef] [PubMed]

32. Hama, T.; Yamamoto, K.; Yaguchi, Y.; Murakami, D.; Sasaki, H.; Yamato, M.; Okano, T.; Kojima, H. Autologous human nasal epithelial cell sheet using temperature-responsive culture insert for transplantation after middle ear surgery. *J. Tissue Eng. Regener. Med.* **2017**, *11*, 1089–1096. [CrossRef] [PubMed]

33. Tang, Z.; Okano, T. Recent development of temperature-responsive surfaces and their application for cell sheet engineering. *Regener. Biomater.* **2014**, *1*, 91–102. [CrossRef] [PubMed]

34. Monteiro, N.; Smith, E.E.; Angstadt, S.; Zhang, W.; Khademhosseini, A.; Yelick, P.C. Dental cell sheet biomimetic tooth bud model. *Biomaterials* **2016**, *106*, 167–179. [CrossRef] [PubMed]

35. Uskokovic, V. Amelogenin in Enamel Tissue Engineering. *Adv. Exp. Med. Biol.* **2015**, *881*, 237–254. [CrossRef] [PubMed]

36. Nanci, A.; Zalzal, S.; Lavoie, P.; Kunikata, M.; Chen, W.; Krebsbach, P.H.; Yamada, Y.; Hammarstrom, L.; Simmer, J.P.; Fincham, A.G.; et al. Comparative immunochemical analyses of the developmental expression and distribution of ameloblastin and amelogenin in rat incisors. *J. Histochem. Cytochem.* **1998**, *46*, 911–934. [CrossRef] [PubMed]

37. Ravindranath, H.H.; Chen, L.S.; Zeichner-David, M.; Ishima, R.; Ravindranath, R.M. Interaction between the enamel matrix proteins amelogenin and ameloblastin. *Biochem. Biophys. Res. Commun.* **2004**, *323*, 1075–1083. [CrossRef] [PubMed]

38. Fukumoto, S.; Kiba, T.; Hall, B.; Iehara, N.; Nakamura, T.; Longenecker, G.; Krebsbach, P.H.; Nanci, A.; Kulkarni, A.B.; Yamada, Y. Ameloblastin is a cell adhesion molecule required for maintaining the differentiation state of ameloblasts. *J. Cell Biol.* **2004**, *167*, 973–983. [CrossRef] [PubMed]

39. Fukumoto, S.; Yamada, A.; Nonaka, K.; Yamada, Y. Essential roles of ameloblastin in maintaining ameloblast differentiation and enamel formation. *Cells Tissues Organs* **2005**, *181*, 189–195. [CrossRef] [PubMed]

40. Hatakeyama, J.; Fukumoto, S.; Nakamura, T.; Haruyama, N.; Suzuki, S.; Hatakeyama, Y.; Shum, L.; Gibson, C.W.; Yamada, Y.; Kulkarni, A.B. Synergistic roles of amelogenin and ameloblastin. *J. Dent. Res.* **2009**, *88*, 318–322. [CrossRef] [PubMed]

41. Deutsch, D.; Palmon, A.; Fisher, L.W.; Kolodny, N.; Termine, J.D.; Young, M.F. Sequencing of bovine enamelin ("tuftelin") a novel acidic enamel protein. *J. Biol. Chem.* **1991**, *266*, 16021–16028. [PubMed]

42. Martens, W.; Bronckaers, A.; Politis, C.; Jacobs, R.; Lambrichts, I. Dental stem cells and their promising role in neural regeneration: An update. *Clin. Oral Investig.* **2013**, *17*, 1969–1983. [CrossRef] [PubMed]

43. Chen, S.; Gluhak-Heinrich, J.; Wang, Y.H.; Wu, Y.M.; Chuang, H.H.; Chen, L.; Yuan, G.H.; Dong, J.; Gay, I.; MacDougall, M. Runx2, osx, and dspp in tooth development. *J. Dent. Res.* **2009**, *88*, 904–909. [CrossRef] [PubMed]

44. Yurchenco, P.D.; O'Rear, J.J. Basal lamina assembly. *Curr. Opin. Cell Biol.* **1994**, *6*, 674–681. [CrossRef]

45. Nagai, N.; Nakano, K.; Sado, Y.; Naito, I.; Gunduz, M.; Tsujigiwa, H.; Nagatsuka, H.; Ninomiya, Y.; Siar, C.H. Localization of type IV collagen a 1 to a 6 chains in basement membrane during mouse molar germ development. *Int. J. Dev. Biol.* **2001**, *45*, 827–831. [PubMed]

46. Simon-Assmann, P.; Bouziges, F.; Arnold, C.; Haffen, K.; Kedinger, M. Epithelial-mesenchymal interactions in the production of basement membrane components in the gut. *Development* **1988**, *102*, 339–347. [PubMed]

47. Simon-Assmann, P.; Spenle, C.; Lefebvre, O.; Kedinger, M. The role of the basement membrane as a modulator of intestinal epithelial-mesenchymal interactions. *Prog. Mol. Biol. Transl. Sci.* **2010**, *96*, 175–206. [CrossRef] [PubMed]
48. Khoshnoodi, J.; Pedchenko, V.; Hudson, B.G. Mammalian collagen IV. *Microsc. Res. Tech.* **2008**, *71*, 357–370. [CrossRef] [PubMed]
49. McGuire, J.D.; Gorski, J.P.; Dusevich, V.; Wang, Y.; Walker, M.P. Type IV collagen is a novel DEJ biomarker that is reduced by radiotherapy. *J. Dent. Res.* **2014**, *93*, 1028–1034. [CrossRef] [PubMed]

nanomaterials

MDPI

Article

Biofunctionalized Lysophosphatidic Acid/Silk Fibroin Film for Cornea Endothelial Cell Regeneration

Joo Hee Choi [1,2], Hayan Jeon [1,2], Jeong Eun Song [1,2], Joaquim Miguel Oliveira [3,4], Rui Luis Reis [3,4] and Gilson Khang [1,2,*]

[1] Department of BIN Convergence Technology, Deokjin-gu, Jeonju-si, Jeollabuk-do 54896, Korea; zooheechoi@jbnu.ac.kr (J.H.C.); wjsgkdis@hanmail.net (H.J.); songje@jbnu.ac.kr (J.E.S.)
[2] Department of Polymer Nano Science & Technology and Polymer BIN Research Center, Chonbuk National University, Deokjin-gu, Jeonju-si, Jeollabuk-do 54896, Korea
[3] 3B's Research Group—Biomaterials, Biodegradables and Biomimetics, University of Minho, Headquarters of the European Institute of Excellence on Tissue Engineering and Regenerative Medicine, AvePark—Parque de Ciência e Tecnologia, Zona Industrial de Gandra, 4805-017 Barco, Guimarães, Portugal; miguel.oliveira@dep.uminho.pt (J.M.O.); rgreis@dep.uminho.pt (R.L.R.)
[4] ICVS/3B's—PT Government Associated Laboratory, Braga/Guimarães, Portugal; The Discoveries Centre for Regenerative and Precision Medicine, Headquarters at University of Minho, Avepark, 4805-017 Barco, Guimarães, Portugal
*   Correspondence: gskhang@jbnu.ac.kr; Tel.: +82-63-270-2355

Received: 28 March 2018; Accepted: 25 April 2018; Published: 30 April 2018

✓ check for updates

Abstract: Cornea endothelial cells (CEnCs) tissue engineering is a great challenge to repair diseased or damaged CEnCs and require an appropriate biomaterial to support cell proliferation and differentiation. Biomaterials for CEnCs tissue engineering require biocompatibility, tunable biodegradability, transparency, and suitable mechanical properties. Silk fibroin-based film (SF) is known to meet these factors, but construction of functionalized graft for bioengineering of cornea is still a challenge. Herein, lysophosphatidic acid (LPA) is used to maintain and increase the specific function of CEnCs. The LPA and SF composite film (LPA/SF) was fabricated in this study. Mechanical properties and in vitro studies were performed using a rabbit model to demonstrate the characters of LPA/SF. ATR-FTIR was characterized to identify chemical composition of the films. The morphological and physical properties were performed by SEM, AFM, transparency, and contact angle. Initial cell density and MTT were performed for adhesion and cell viability in the SF and LPA/SF film. Reverse transcription polymerase chain reactions (RT-PCR) and immunofluorescence were performed to examine gene and protein expression. The results showed that films were designed appropriately for CEnCs delivery. Compared to pristine SF, LPA/SF showed higher biocompatibility, cell viability, and expression of CEnCs specific genes and proteins. These indicate that LPA/SF, a new biomaterial, offers potential benefits for CEnCs tissue engineering for regeneration.

Keywords: cornea endothelial cells; tissue engineering; regeneration; silk fibroin; lysophosphatidic acid

1. Introduction

Cornea is the outer layer of the eye and has three individual layers: the epithelium, stroma, and endothelium [1]. The corneal endothelium is a barrier for metabolic activity that plays an important role in maintaining transparency by utilizing an ATPase pump [2]. Damaged or diseased corneal endothelial cells (CEnCs) are difficult to regenerate due to break down of G1-cell cycle phase. Loss of cell density which is caused by expansion of CEnCs cell size rather than proliferation is also a factor in degeneration of CEnCs. A loss of CEnCs result in corneal edema, blindness, visual acuity, etc. [3,4]. There are nearly 10 million cases of worldwide vision loss due to corneal blindness [5,6]. Therefore, cornea transplantation or replacement of the endothelial cell layer is needed [3]. However, it is reported that although the corneal graft rejection rate is less than 10%, the immunological rejection rate increases to 25% after 4–5 years of implantation and continuously increases over time [6–8]. Thus, the development of tissue engineering strategies for an effective alternative to conventional corneal grafts is increasing, and new biomaterials are required for the development of corneal replacement n. The important factors that should be considered for bioengineered corneas are transparency, biodegradability, biocompatibility, water permeability, possession of essential nutrients for CEnCs, and appropriate mechanical properties for ease handling [2].

Bombyx mori silk is a structural protein and is widely used in tissue engineering biomaterials due to its biocompatibility, biodegradation, tunable mechanical properties, and non-immunogenic response in vivo [12–15]. Silk substrates have been shown to support cell adhesion, mobility, proliferation, and differentiation by mimicking extracellular matrix (ECM) [16–20]. The silk material is made in variety forms such as sponge, hydrogel, fiber, and film [16,17,19–27]. The most often-used structure for corneal tissue engineering is film [2,3,15,28–33]. However, studies are still preceding to improve silk material properties to enhance cell function, proliferation, and biocompatibility [3,12,15,30,32]. Silk fibroin in film form can be incorporated with bio-functional molecules and other biomaterials to generate functional matrices. Also, patterned silk film surfaces can produce high-resolution surface features [21–35].

Herein, lysophosphatidic acid-incorporated silk fibroin film (LPA/SF) was designed for an efficient bioengineering of CEnCs graft. LPA is an endogenous glycerophospholipid signaling molecule, a ligand activator, and has been reported to stimulate growth of fibroblasts, keratinocytes, and endothelial cells, and plays many biological functions in the nervous system [36,37]. The LPA is a critical serum component and affects cell attachment, proliferation, migration, and viability [38]. Epithelial cells, fibroblasts, or platelets are reported to release LPA at injured or inflammation sites [38–40]. Also, LPA is released in the injured cornea, and it is predicted that these factors support cell migration and proliferation to regenerate a wound [38].

The LPA/SF was characterized both mechanically and biochemically to determine suitability of CEnCs carrier.

2. Materials and Methods

2.1. Preparation of SF Solution

Silkworm cocoons were cut into the pieces, and 10 g of silkworm cocoons were added into a boiling 0.02 M Na_2CO_3 (Showa Chemical, Tokyo, Japan) in distilled water (DW) for 30 min to remove sericin. After 30 min, silkworm cocoons were washed with DW and fully dried in 60 °C oven for overnight. The dried SF was dissolved in a 9.3M LiBr (Kanto chemical, Tokyo, Japan) for 4 h in the 60 °C to prepare SF solution. The LiBr was removed via dialysis of SF solution in a dialysis tube (SnakeSkin®Dialysis Tubing 3500 MWCO, Thermo Fisher Scientific, Waltham, MA, USA) for 48 h, and the SF aqueous solution was kept at 4 °C until usage.

2.2. Fabrication of LPA/SF

Pure 8% SF aqueous solution was used in this study. The LPA (Sigma-Aldrich, St. Louis, MO, USA) was dissolved in chloroform:methnol:acetic acid 95:5:5 and incorporated in SF aqueous solution to make the final concentration of 20 µM LPA/SF. The SF and LPA/SF solution were transferred to petri dish to make a thickness of 6–8 µm films and dried under a clean bench to avoid contamination. The films were cross-linked with methanol for 30 min at room temperature. For further sterilization, the films were treated with 70% ethanol under a UV light for 30 min and washed 3 times with PBS for 20 min.

2.3. Characterizations

The chemical structure of SF and LPA/SF was analyzed using ATR-FTIR (Perkin Elmer, Waltham, MA, USA) at the spectra wavelength range of 4000–400 cm^{-1}.

The morphology was investigated by field emission scanning electron microscopy (FESEM, Hitachi S4700) and atomic force microscopy (AFM) on a Scanning Probe Microscope XE 70 (Multimode-8, Bruker, Billerica, MA, USA). Contact angle characterization was carried out by employing water contact goniometer (TantecTM, CAM-PLUS Micro, Schaumburg, IL, USA) to measure the hydrophilicity of tissue culture polystyrene (TCP), SF, and LPA/SF. Transparency of SF and LPA/SF was evaluated by SYNERGY Mx spectrophotometer (BioTek, Winooski, VT, USA) at a wavelength range of 380 nm–780 nm after immersing in PBS.

2.4. Isolation and Culture of Rabbit CEnCs (rCEnCs)

rCEnCs were cultured in endothelial growth medium-2 (EGM-2, Lonza, Walkersville, MD, USA) supplemented with 10% fetal bovine serum (FBS, Gibco, Big Cabin, OK, USA) and 1% penicillin/streptomycin (PS, Gibco, USA), epidermal growth factor (EGF), vascular endothelial growth factor (VEGF), fibroblast growth factor (FGF), insulin-like growth factor (IGF), hydrocortisone, gentamicin, and amphotericin-B under standard culture conditions (5% CO_2, and 37 °C). rCEnCs were isolated from New Zealand white rabbits (4 weeks old, Female). Animal experiment procedures were approved by Chonbuk National University Animal Care Committee, Jeonju, South Korea. Briefly, eyeballs were removed from rabbits and transferred to PBS. The eyeballs were washed 3 times with PBS under a clean bench. Soft tissues were removed, and cornea was cut from the eyeball. The endothelium with Descemet membrane was stripped from stroma. rCEnCs were transferred into a 0.2% collagenase A (Roche, Germany) and digested for 40 min in an incubator with the condition of 5% CO_2, and 37 °C. After digestion, the solution was centrifuged at 1500 rpm for 5 min at 4 °C. The rCEnCs pallet was suspended in the culture media and incubated under standard conditions (5% CO_2, and 37 °C). The media was changed every 2 days, and passage 0 of primary rCEnCs was used for this study.

2.5. Morphology Analysis

The rCEnCs were seeded on the SF and LPA/SF at density of 100 cells/mm^2 and cultured in EGM-2 medium for 5 days. The culture media was changed every 2 days. The media solution was removed, and the films were washed with PBS. The films with adhered cells were fixed with 2.5% glutaraldehyde (Sigma-Aldrich, USA) and different concentrations of ethanol (50%, 60%, 70%, 80%, 90%, and 100%) were used sequentially to dehydrate the films. The ethanol was changed every 20 min for dehydration and dried for 24 h in room temperature before FESEM evaluation.

2.6. Initial Attachment

The density of 500 cells/mm^2 was seeded on TCP, SF, and LPA/SF in the culture media. After 30 min of culture, the medium was aspirated and fixed with a cold methanol at 4 °C for 24 h. The samples were washed with PBS and stained with DAPI (Santa Cruz Biotechnology, Santa Cruz, CA, USA). The initial attachment of rCEnCs on the TCP, and films were investigated by fluorescence microscopy (Nikon Eclipse TE-2000U, Nikon, Tokyo, Japan), and the cell number was counted using the Image J program ($n = 3$).

2.7. Cell Viability

MTT (3-[4,-dimethylthiazol-2-yl]-2,5-diphenyltetrazoliumbromide;thiazolyl blue, Sigma-Aldrich, USA) assay was performed for the rCEnC viability on the SF and LPA/SF. The 100 cells/mm^2 per well were seeded on TCP, SF, and LPA/SF and cultured in EGM-2. The rCEnCs were cultured for 1, 3, and 7 days, and the culture media was changed every 2 days. The samples were replaced with fresh culture medium with the addition of MTT solution (50 mg/mL in PBS) to make 10% MTT of medium volume and incubated under standard conditions (5% CO_2, and 37 °C) for formazan crystal formation. After 4 h of incubation, the solution was aspirated and 1 mL of dimethyl sulfoxide (DMSO) was added to dissolve formazan crystal. Finally, the absorbance of the solution was evaluated at 570 nm using microplate reader (Synergy MX, Biotek, Vernusky, VT, USA) ($n = 3$).

2.8. mRNAs Expression

The rCEnCs seeded with the density of 100 cells/mm^2 films and TCP were cultured for 3 and 5 days. Trizol (Invitrogen, Carlsbad, CA, USA) and chloroform (Sigma-Aldrich, USA) were used to extract mRNA and centrifuged at 12,000 rpm in 4 °C for 15 min. The supernatant was transferred to a 1.5 mL Eppendorf tube. Iso-propanol (Sigma-Aldrich, USA) was added and kept in 4 °C overnight. Isolated mRNA was dissolved in RNase-DNase free water (Gibco, USA). The gene markers of voltage-dependent anion-selective channel 2 (VDAC2), voltage-dependent anion-selective channel 3 (VDAC3), chloride channel protein 2 (CLCN2), and sodium/bicarbonate co-transporter (NBC1) were evaluated and normalized using β-actin a housekeeping gene. Gene expression was measured by electrophoresis on 1% (w/v) agarose gel containing Ethidium Bromide (EtBr, Sigma-Aldrich, USA). Images were obtained under a UV light (FluorChem FC2, Alpha Innotech, San Leandro, USA) at 360 nm.

2.9. Immunohistological Analysis

The expression of Na^+/K^+-ATPase and ZO-1 were measured to identify rCEnCs on SF and LPA/SF. The histological expression was evaluated after 3 days of culture. The rCEnCs were fixed with 4% formaldehyde (Sigma-Aldrich, USA) at 4 °C overnight and washed with PBS three times. A protein-blocking solution (DAKO, Glostrup, Denmark) was added for 15 min at room temperature to prevent non-specific binding. Fixed samples were incubated with the primary antibodies anti- Na^+/K^+-ATPase and anti-ZO-1 (1:200, Sata Cruz Biotechnology, Dallas, TX, USA) at 4 °C for overnight. Alexa Fluor®594-conjugated AffiniPure Donkey Anti-Rabbit IgG (1:300, Jackson Immuno Research Laboratories, Inc., West Baltimore Pike West Grove, PA, USA) was used as a secondary antibody. The images were taken by confocal laser scanning microscope (LSM 510 META, Zeiss, Oberkochen, Germany) installed in the Center for University-Wide Research Facilities (CURF) at Chonbuk National University.

3. Results

3.1. Characterization of SF and LPA/SF

3.1.1. ATR-FTIR Spectroscopy

The LPA, SF, and LPA/SF components and structure confirmation were analyzed by ATR-FTIR. The crystallized SF is shown in the 1700 cm^{-1}–1200 cm^{-1} range. The β-sheet formation of the film is displayed by the three amide peaks. The C=O stretch, which is amide I (2), is shown at 1630 cm^{-1}–1650 cm^{-1}; amide II (3), which is N-H band, appears at 1520 cm^{-1}–1540 cm^{-1} and C=N stretch the amide III (4) is shown at the range of 1230 cm^{-1}–1270 cm^{-1}. Also, the –OH peak (1) was shown at the range of 3000 cm^{-1}–3650 cm^{-1}. The result of LPA/SF shows deeper depth transmittance than the SF. The LPA/SF peak showed slightly different peak at the wavelength range of 3100 cm^{-1}–2910 cm^{-1} compared to the SF (Figure 1).

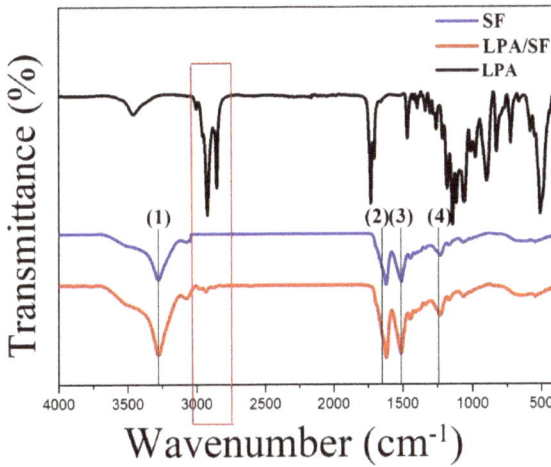

Figure 1. ATR-FTIR spectroscopy of SF and LPA/SF spectra wavelength rage of 4000 cm^{-1}–400 cm^{-1}.

3.1.2. Transparency

The transparency of the films was studied by spectrophotometer at the wavelength range of 380 nm–780 nm. The gross image of SF and LPA/SF and the transparency of the fabricated films and TCP in the visible range is shown (Figure 2a,b). TCP, which is a commercially available material, was set as a positive control. The SF showed the highest transparency, but there was no significant difference between LPA/SF and TCP. The transparency of films with cell culture displayed LPA/SF slightly more transparently than TCP and the pristine SF.

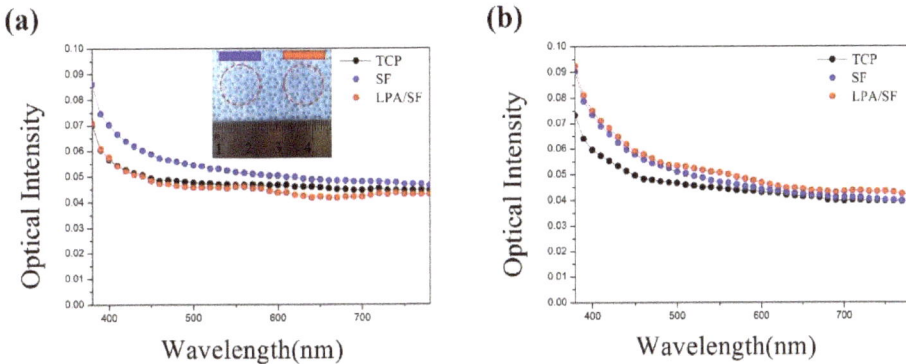

Figure 2. Gross image and transparency of SF (red) and LPA/SF (blue) at wavelength of 380 nm 780 nm (**a**) without cell culture and (**b**) with cell culture (*n* = 3).

3.1.3. Hydrophilicity

The contact angle of a water droplet on SF and LPA/SF was measured for 10 min. The difference of contact angle between SF and LPA/SF was not significant for 5 min. However, LPA/SF showed lower contact angle than the SF after 6 min (Figure 3).

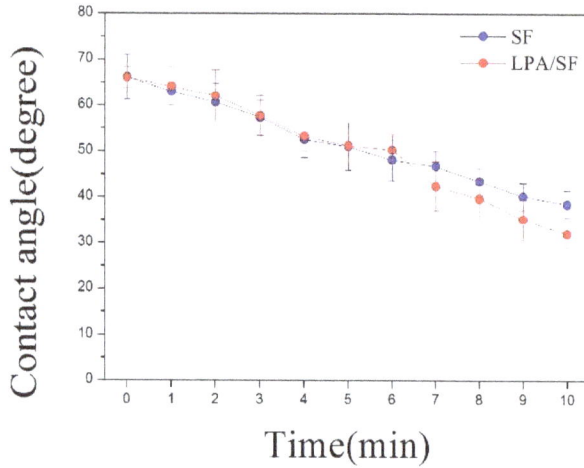

Figure 3. Contact angle of single water droplet (2 μL) on SF and 20 μM LPA/SF observed for 10 min (*n* = 5).

3.1.4. Surface Morphology and Roughness

The surface properties of the SF and LPA/SF were studied by FESEM and three-dimensional (3D) AFM images. There was no significant difference between each film on the FESEM image (Figure 4a). However, the topographic AFM images of bare SF and LPA/SF differed (Figure 4b). LPA/SF showed rougher surface when compared to the SF.

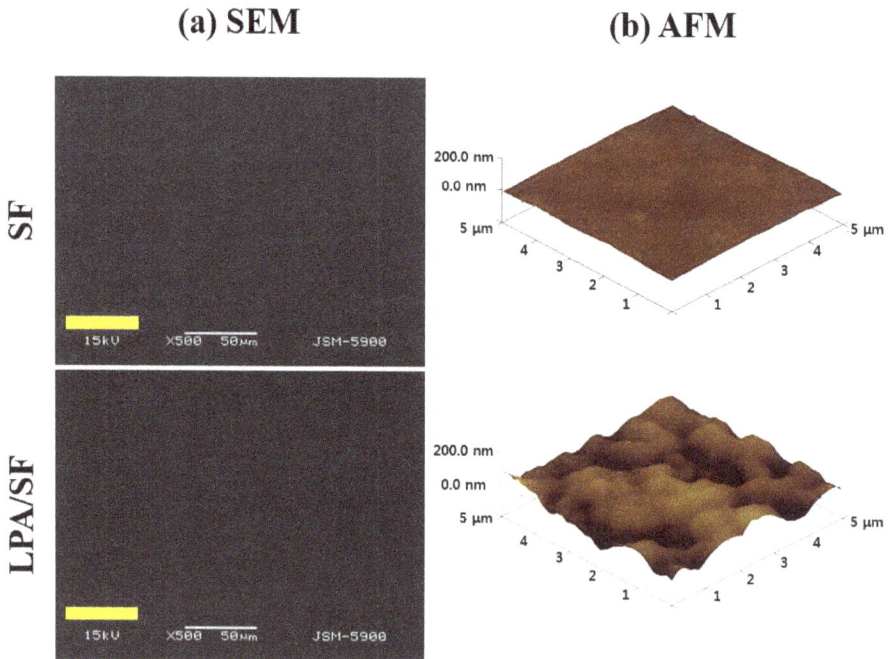

Figure 4. (**a**) FESEM and (**b**) AFM images of as-fabricated SF and LPA/SF.

3.2. In Vitro Study

3.2.1. Monolayer Formation

The morphology of rCEnCs cultured on films was evaluated by FESEM. The morphology, cell density, and ECM secretion of rCEnCs were significantly different between SF and LPA/SF (Figure 5a,b). The hexagonal shape of cells (red line), which is the basic structure of the endothelial cells, was displayed in both films, but the ECM secretion between cells was significantly higher in LPA/SF (yellow arrow).

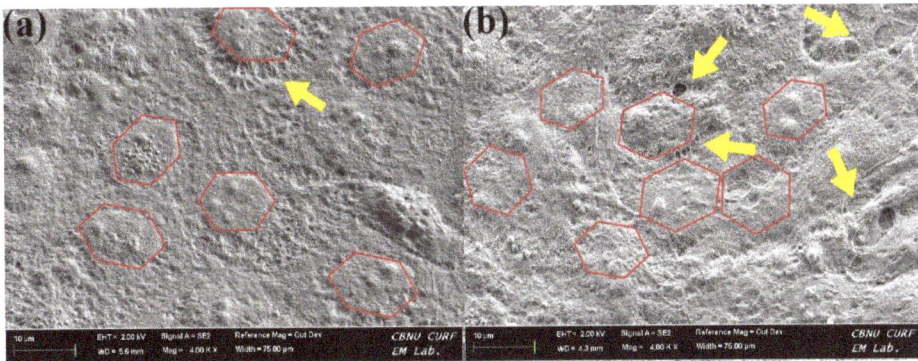

Figure 5. FESEM images of 5 days-cultured rCEnCs morphology on (**a**) SF and (**b**) LPA/SF.

3.2.2. Initial Attachment

The fluorescence image of TCP, SF and LPA/SF displayed different cell attachment. TCP showed the highest cells and LPA/SF presented analogous cell density to TCP (Figure 6a). The initial cell attachment on TCP the positive control was the highest (683.15 ± 42.66 cells/mm^2). The LPA/SF displayed a similar level of TCP initial cell attachment (575.1 ± 123.34 cells/mm^2). The SF cell density was the lowest (308.7 ± 63.12 cells/mm^2) (Figure 6b).

(a)

(b)

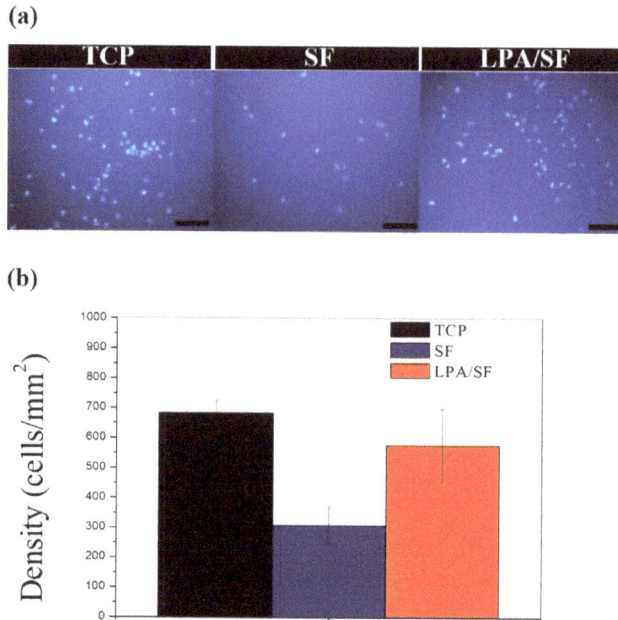

Figure 6. Initial attachment of rCEnCs on TCP, SF, and LPA/SF. (**a**) DAPI staining for initial attachment evaluation (scale bar 100 μm) and (**b**) initial attachment of rCEnCs on the films (*n* = 3).

3.2.3. Cell Proliferation

TCP showed the highest cell proliferation for 7 days. Compared to bare SF, LPA/SF was found to increase in cellular number. The LPA/SF displayed higher cell density for 7 days, which implies better rCEnCs growth than the SF (Figure 7).

3.2.4. Specific mRNAs Expression

mRNAs expression was studied by RT-PCR using CEnCs-associated genes such as VDAC2, VDAC3, NBC1, and CLCN2. β-Actin housekeeping gene was used for normalization. TCP was set as a positive control. LPA/SF showed more enhanced gene expression for 5 days compared to SF alone (Figure 8).

3.2.5. Immunohistological Evaluation

The SF and LPA/SF cultured with rCEnCs were stained with Na^+/K^+-ATPase (Na-K), which is related to cornea transparency and tight junction protein ZO-1. Both films were properly expressed without any remarkable changes of rCEnCs morphology (Figure 9a). The LPA/SF showed higher fluorescence intensity, which was measured by the Image J program. (Figure 9b).

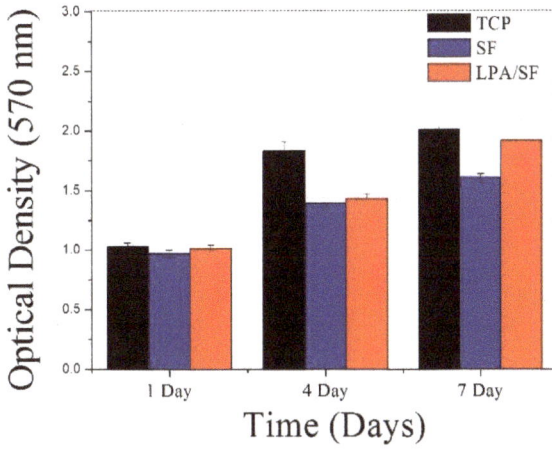

Figure 7. Proliferation assay of cell-cultured TCP, SF, and LPA/SF in EGM-2 (*n* = 3).

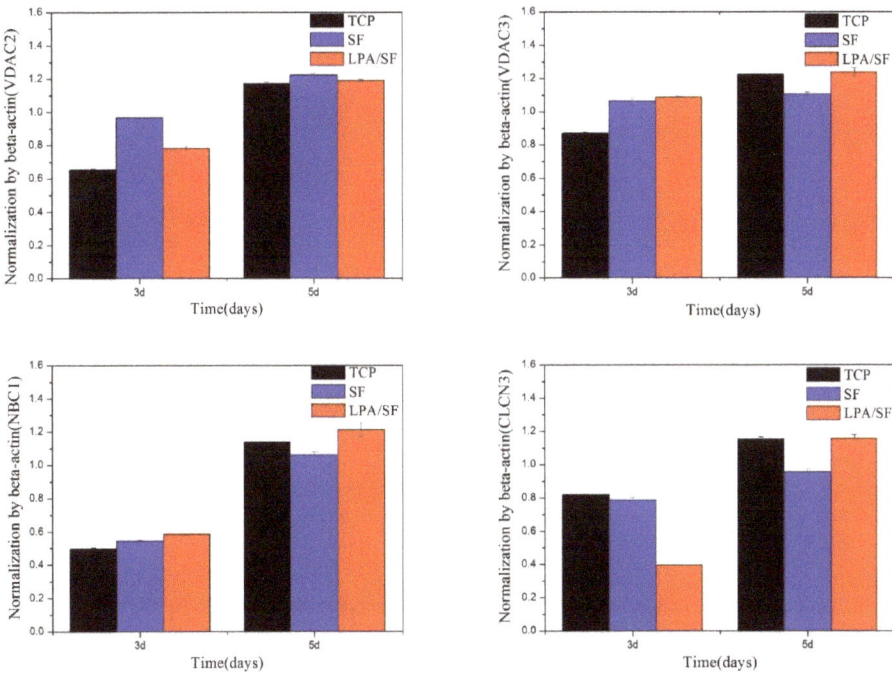

Figure 8. CEnCs-specific gene expressions of different films with cell culture by RT-PCR, normalized by β-actin (*n* = 3).

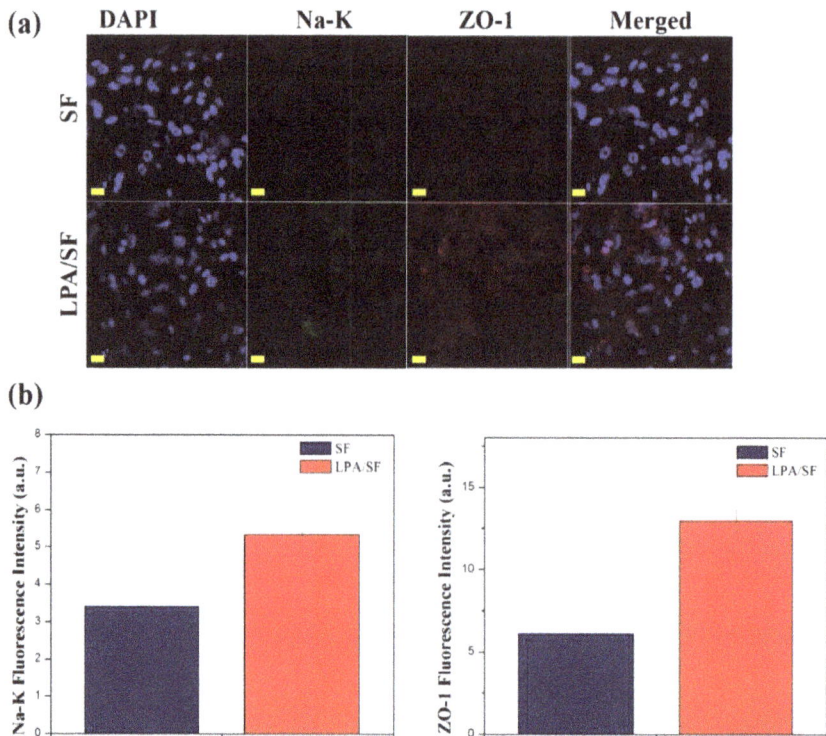

Figure 9. Immunofluorescence staining images of Na-K and ZO-1 of rCEnCs cultured on SF and LPA/SF (Scale bar 50 µm). (**a**) Fluorescence intensity of Na-K and ZO-1 analyzed by image J program (**b**).

4. Discussion

The structure of SF and LPA/SF was confirmed by the ATR-FTIR spectroscopy. Both SF and LPA/SF displayed proper crystallized Silk fibroin [15]. The LPA/SF presented a small peak at the wavelength rage of 3100 cm^{-1}–2910 cm^{-1}, which is similar peak to the one shown in the LPA. Moreover, the LPA/SF exhibited deeper depth of the absorption band compared to SF. It is speculated that the increase in intensity may be due to the overlap of the LPA and silk fibroin peak. However, the concentration of the LPA in the SF was low that there was no particular difference.

Optically transparent SF and LPA SF were formed (Figure 2). The transparency is important in cornea tissue engineering to provide a clear vision in vivo and monitor cell behavior, process of healing, and check on any sign of infection [15,41]. Notably, the optical intensity of healthy human acellular corneal stroma is 0.1–0.13 at the wavelength of 380 nm–780 nm [2,42,43]. Figure 2 shows the as-fabricated films transparency at the wavelength range of 380 nm–780 nm. The pristine SF showed higher transparency than the LPA/SF. However, when films were cultured with rCEnCs, LPA/SF showed slightly higher transparency. This suggests that LPA/SF has the potential to be used as a corneal substitute.

Hydrophilicity property is important, because it effects cell attachment, proliferation, and migration on the substrate [28,29,35]. The measurement of contact angle is utilized to determine the hydrophilicity and applicability of the films. The hydrophilicity was evaluated at 0 min and the hydrophobicity was analyzed at 10 min to characterize the hydrophilicity/hydrophobicity. LPA-incorporated SF showed higher hydrophilicity than the bare SF (Figure 3). Hydrophilic film provides essential nutrients and prevents loss of body fluids [44–46].

The surface characteristics of film for CEnC regeneration is crucial for interaction between cells and proliferation [47]. The surface was identified by the FESEM and AFM. The FESEM result showed a smooth surface in both films (Figure 4a). However, the topography of the films was significantly different in AFM, which measures the surface roughness in nanoscale [48]. The roughness was significantly high in LPA/SF, whereas the SF presented relatively smooth surface (Figure 4b). It is reported that a rough surface provides greater cell attachment and a greater proliferation environment [48]. The result of the cell morphology on the films showed concomitant results of foregoing discussion. The films cultured with rCEnCs displayed monolayer on all films and presented hexagonal shape of rCEnCs. However, large amount of ECM release between rCEnCs in LPA/SF was exhibited (Figure 5b). Interaction between ECM and single layer of rCEnCs with the specific hexagonal shape regulates the hydration of corneal stroma and controls transparency of the cornea in the anterior chamber of eyeball [30,49,50].

The initial attachment to a graft, which defines affinity toward the specific substrate, is an important factor to consider in tissue engineering [28,32,35]. A high rate of initial attachment signifies that less time is required to reach the demanded confluency to the target graft and higher cell proliferation [15,43,45]. The evaluation of the initial cell attachment to the films showed that the LPA/SF was significantly higher than SF and similar rate to the positive control TCP (Figure 6a,b). The viability and proliferation of CEnCs is important for vision recovery when transplanted [34]. The density of CEnCs must reach around 500 cells/mm^2, otherwise it cannot function properly and causes edema [15,51]. Both films increased significantly over 4 days compared to 1 day. However, the difference was observed after 7 days, at which point LPA/SF displayed significantly increased rate of proliferation as similar to the positive control TCP (Figure 7). It is predicted that LPA activated ligand of cells and led to cell proliferation [37]. Furthermore, the TCP is widely known as the best material for tissue culture of cell adhesion and proliferation [52]. However, TCP cannot be transplantable. The fact that LPA/SF results in analogous level of proliferation to the TCP suggests that LPA/SF is a beneficial material for CEnCs tissue engineering. This is because LPA/SF can be transplantable, and carries culture property similar to the TCP.

For further confirmation, specific mRNAs and functional protein expressions were obtained. VDAC2 and VDAC3 are known to regulate the interaction among cells, proteins, and small molecules [53]. CLCN2 controls pH, transports organic molecules, and effects cells migration, proliferation, and differentiation [34,54]. NBC1 is one of the NBC proteins and involves in absorption and secretion of cellular HCO_3^- and also intracellular pH regulation [54]. LPA/SF cultured with rCEnCs showed more gene expression than SF on VDAC3, CLCN2, and NBC1 and slightly lower expression on VDAC2 on 7 days (Figure 8). Considering the fact that LPA is involved in epithelial growth, it is expected that LPA increases the favorable environment for cell growth and genes expression of SF [37]. Moreover, the LPA/SF displaying the similar level of gene expression to the positive control TCP is significant. This fact indicates that the LPA/SF can provide proper environment similar to the TCP. The protein expressions results showed that rCEnCs on all films were displayed properly without any remarkable changes in the cells morphology (Figure 9). Na-K is the typical CEnCs marker which is responsible in clearing water from the corneal stroma and providing transparency [2]. The LPA/SF exhibited higher expression of Na-K compared to the SF (Figure 9a,b). It is assumed that the LPA supported the SF to provide enhanced environment for rCEnCs functionality. The LPA/SF also showed higher expression level in ZO-1 which is a tight junction protein. This may be attributed to LPA considering the fact that LPA is known to stimulate cells growth and biological function [3,36–40]. In conclusion, LPA/SF provided similar condition to positive control TCP and higher CEnCs specific gene and protein expression. It is envisioned that LPA/SF can be a promising substrate for CEnCs delivery.

5. Conclusions

In this study, proper and enhanced film for corneal tissue regeneration was designed. Both SF and LPA/SF showed appropriate mechanical properties and enhanced biological properties in vitro. rCEnCs successfully adhered, proliferated, and expressed CEnCs specific genes and proteins on the both films. Further work is required to obtain an insight into how cells are modulated and respond to LPA/SF. Considering the overall results, it is expected that LPA/SF can be employed to enhance the clinical efficacy of the delivery of CEnCs to regenerate diseased or damaged CEnCs.

Author Contributions: H.J. and J.C. designed the experiments; H.J. and J.C. performed the experiments; J.C., H.J., and J.S. analyzed the data; J.O., R.R., and G.K. contributed reagents/materials/analysis tools; J.C. wrote the paper.

Acknowledgments: This research was supported by Basic Science Research Program through the National Research Foundation of Korea (NRF) funded by the Ministry of Science, ICT, & Future Planning (NRF-2017R1A2B3010270). This research was supported by a grant of the Korea Health Technology R&D Project through the Korea Health Industry Development Institute (KHIDI), funded by the Ministry of Health &Welfare, Republic of Korea (grant number: HI15C2996).

Conflicts of Interest: The authors declare no conflict of interest.

References

1. Wang, S.; Ghezzi, C.E.; Gomes, R.; Pollard, R.E.; Funderburgh, J.L.; Kaplan, D.L. In vitro 3D corneal tissue model with epithelium, stroma, and innervation. *Biomaterials* **2017**, *112*, 1–9. [CrossRef] [PubMed]
2. Kim, D.K.; Sim, B.R.; Kim, J.I.; Khang, G. Functionalized silk fibroin film scaffold using β-Carotene for cornea endothelial cell regeneration. *Colloids Surfaces B Biointerfaces* **2018**, *164*, 340–346. [CrossRef] [PubMed]
3. Kim, E.Y.; Tripathy, N.; Cho, S.A.; Joo, C.K.; Lee, D.; Khang, G. Bioengineered neo-corneal endothelium using collagen type-I coated silk fibroin film. *Colloids Surfaces B Biointerfaces* **2015**, *136*, 394–401. [CrossRef] [PubMed]
4. Mimura, T.; Yamagami, S.; Amano, S. Corneal endothelial regeneration and tissue engineering. *Prog. Retin. Eye Res.* **2013**, *35*, 1–17. [CrossRef] [PubMed]
5. Aldave, A.J.; Dematteo, J.; Glasser, D.B.; Tu, E.Y.; Iliakis, B.; Nordlund, M.L.; Misko, J.; Verdier, D.D.; Yu, F. Report of the eye bank association of America medical advisory board subcommittee on fungal infection after corneal transplantation. *Cornea* **2013**, *32*, 149–154. [CrossRef] [PubMed]
6. Lawrence, B.D.; Marchant, J.K.; Pindrus, M.A.; Omenetto, F.G.; Kaplan, D.L. Silk film biomaterials for cornea tissue engineering. *Biomaterials* **2009**, *30*, 1299–1308. [CrossRef] [PubMed]
7. Myung, D.; Koh, W.; Bakri, A.; Zhang, F.; Marshall, A.; Ko, J.; Noolandi, J.; Carrasco, M.; Cochran, J.R.; Frank, C.W.; et al. Design and fabrication of an artificial cornea based on a photolithographically patterned hydrogel construct. *Biomed. Microdevices* **2007**, *9*, 911–922. [CrossRef] [PubMed]
8. Ilhan-Sarac, O.; Akpek, E.K. Current concepts and techniques in keratoprosthesis. *Curr. Opin. Ophthalmol.* **2005**, *16*, 246–250. [CrossRef] [PubMed]
9. Liu, W.; Merrett, K.; Griffith, M.; Fagerholm, P.; Dravida, S.; Heyne, B.; Scaiano, J.C.; Watsky, M.A.; Shinozaki, N.; Lagali, N.; et al. Recombinant human collagen for tissue engineered corneal substitutes. *Biomaterials* **2008**, *29*, 1147–1158. [CrossRef] [PubMed]
10. Alaminos, M.; Sánchez-Quevedo, M.D.C.; Muñoz-Ávila, J.I.; Serrano, D.; Medialdea, S.; Carreras, I.; Campos, A. Construction of a complete rabbit cornea substitute using a fibrin-agarose scaffold. *Investig. Ophthalmol. Vis. Sci.* **2006**, *47*, 3311–3317. [CrossRef] [PubMed]
11. Hu, X.; Lui, W.; Cui, L.; Wang, M.; Cao, Y. Tissue engineering of nearly transparent corneal stroma. *Tissue Eng.* **2005**, *11*, 1710–1717. [CrossRef] [PubMed]
12. Hazra, S.; Nandi, S.; Naskar, D.; Guha, R.; Chowdhury, S.; Pradhan, N.; Kundu, S.C.; Konar, A. Non-mulberry Silk Fibroin Biomaterial for Corneal Regeneration. *Sci. Rep.* **2016**, *6*. [CrossRef] [PubMed]
13. Kasoju, N.; Bora, U. Silk fibroin in tissue engineering. *Adv. Healthc. Mater.* **2012**, *1*, 393–412. [CrossRef] [PubMed]
14. Moon, B.M.; Kim, D.K.; Park, H.J.; Ju, H.W.; Lee, O.J.; Kim, J.H.; Lee, J.M.; Lee, J.S.; Park, C.H. Fabrication and characterization of three-dimensional silk fibroin scaffolds using a mixture of salt/sucrose. *Macromol. Res.* **2014**, *22*, 1268–1274. [CrossRef]

15. Kim, D.K.; Sim, B.R.; Khang, G. Nature-Derived Aloe Vera Gel Blended Silk Fibroin Film Scaffolds for Cornea Endothelial Cell Regeneration and Transplantation. *ACS Appl. Mater. Interfaces* **2016**, *8*, 15160–15168. [CrossRef] [PubMed]

16. Kim, D.K.; Kim, J.I.; Hwang, T.I.; Sim, B.R.; Khang, G. Bioengineered Osteoinductive Broussonetia Kazinoki/ Silk Fibroin Composite Scaffolds for Bone Tissue Regeneration. *ACS Appl. Mater. Interfaces* **2016**. [CrossRef] [PubMed]

17. Kim, D.K.; In Kim, J.; Sim, B.R.; Khang, G. Bioengineered porous composite curcumin/silk scaffolds for cartilage regeneration. *Mater. Sci. Eng. C* **2017**, *78*, 571–578. [CrossRef] [PubMed]

18. Han, K.-S.; Song, J.E.; Tripathy, N.; Kim, H.; Moon, B.M.; Park, C.H.; Khang, G. Effect of pore sizes of silk scaffolds for cartilage tissue engineering. *Macromol. Res.* **2015**, *23*, 1091–1097. [CrossRef]

19. Sim, B.R.; Kim, H.M.; Kim, S.M.; Kim, D.K.; Song, J.E.; Park, C.-H.; Khang, G. Osteogenesis differentiation of rabbit bone marrow-mesenchymal stem cells in silk scaffold loaded with various ratios of hydroxyapatite. *Polymer* **2016**, *40*, 915–924. [CrossRef]

20. Lee, D.H.; Tripathy, N.; Shin, J.H.; Song, J.E.; Cha, J.G.; Min, K.D.; Park, C.H.; Khang, G. Enhanced osteogenesis of β-tricalcium phosphate reinforced silk fibroin scaffold for bone tissue biofabrication. *Int. J. Biol. Macromol.* **2017**, *95*, 14–23. [CrossRef] [PubMed]

21. Suganya, S.; Venugopal, J.; Ramakrishna, S.; Lakshmi, B.S.; Dev, V.R. Aloe Vera/Silk Fibroin/Hydroxyapatite Incorporated Electrospun Nanofibrous Scaffold for Enhanced Osteogenesis. *J. Biomater. Tissue Eng.* **2014**, *4*, 9–19. [CrossRef]

22. Meinel, L.; Betz, O.; Fajardo, R.; Hofmann, S.; Nazarian, A.; Cory, E.; Hilbe, M.; McCool, J.; Langer, R.; Vunjak-Novakovic, G.; et al. Silk based biomaterials to heal critical sized femur defects. *Bone* **2006**, *39*, 922–931. [CrossRef] [PubMed]

23. Jang, N.K.; Lee, S.E.; Cha, S.R.; Kim, C.H.; Jeong, H.K.; Kim, S.Y.; Shin, J.H.; Song, J.E.; Park, C.H.; Khang, G. Osteogenic Differentiation of Bone Marrow Stem Cell in Silk Fibroin Scaffold. *Int. J. Tissue Regen.* **2015**, *6*, 56–63.

24. Shen, W.; Chen, X.; Hu, Y.; Yin, Z.; Zhu, T.; Hu, J.; Chen, J.; Zheng, Z.; Zhang, W.; Ran, J.; et al. Long-term effects of knitted silk-collagen sponge scaffold on anterior cruciate ligament reconstruction and osteoarthritis prevention. *Biomaterials* **2014**, *35*, 8154–8163. [CrossRef] [PubMed]

25. Shin, J.H.; Lee, D.H.; Song, J.E.; Cho, S.J.; Park, C.H.; Khang, G. Effect of silk sponge concentrations on skin regeneration. *Polymer* **2017**, *41*, 1–6. [CrossRef]

26. Hardy, J.G.; Scheibel, T.R. Composite materials based on silk proteins. *Prog. Polym. Sci.* **2010**, *35*, 1093–1115. [CrossRef]

27. Kim, E.Y.; Tripathy, N.; Park, J.Y.; Lee, S.E.; Joo, C.K.; Khang, G. Silk fibroin film as an efficient carrier for corneal endothelial cells regeneration. *Macromol. Res.* **2015**, *23*, 189–195. [CrossRef]

28. Lee, S.E.; Cha, S.R.; Jang, N.K.; Kim, S.Y.; Kim, E.Y.; Song, J.E.; Park, C.H.; Khang, G. Effect of degumming time of silk films on growth of corneal endothelial cells for tissue engineered endothelialized neo-corneas. *Polymer* **2016**, *40*, 181–187. [CrossRef]

29. Wu, J.; Du, Y.; Watkins, S.C.; Funderburgh, J.L.; Wagner, W.R. The engineering of organized human corneal tissue through the spatial guidance of corneal stromal stem cells. *Biomaterials* **2012**, *33*, 1343–1352. [CrossRef] [PubMed]

30. Sim, B.R.; Kim, D.K.; Jeon, Y.S.; Jeon, H.Y.; Kook, Y.J. Graphene Oxide as an Additive on Silk Fibroin Film for Corneal Endothelial Cell Regeneration. *Int. J. Tissue Regen.* **2016**, *7*, 108–112.

31. Sim, B.R.; Kim, D.K.; Lee, D.H.; Jeon, Y.S.; Tripathy, N. Effect of Quercetin / Silk Fibroin Film on Proliferation of Corneal Endothelial Cells. *Int. J. Tissue Regen.* **2016**, *7*, 1–5.

32. Feng, J.; Mineda, K.; Wu, S.H.; Mashiko, T.; Doi, K.; Kuno, S.; Kinoshita, K.; Kanayama, K.; Asahi, R.; Sunaga, A.; et al. An injectable non-cross-linked hyaluronic-acid gel containing therapeutic spheroids of human adipose-derived stem cells. *Sci. Rep.* **2017**, *7*, 1–13. [CrossRef] [PubMed]

33. Kim, D.K.; Sim, B.R.; Jeon, H.Y.; Song, J.E.; Tripathy, N.; Lee, D.; Khang, G. Silk Fibroin Film Scaffold for Cornea Endothelial Regeneration and Transplantation. *Int. J. Tissue Regen.* **2016**, *7*, 72–76.

34. Kim, S.M.; Kim, E.Y.; Kuk, H.; Kim, S.N.; Sim, B.R.; Kim, D.K.; Joo, C.; Khang, G. A Study for Corneal Endothelial Cell Growth Behavior on Silk Fibroin Substrata. *Int. J. Tissue Regen.* **2015**, *6*, 49–55.

35. Noguchi, K.; Herr, D.; Mutoh, T.; Chun, J. Lysophosphatidic acid (LPA) and its receptors. *Curr. Opin. Pharmacol.* **2009**, *9*, 15–23. [CrossRef] [PubMed]

36. Sugiura, T.; Nakane, S.; Kishimoto, S.; Waku, K.; Yoshioka, Y.; Tokumura, A. Lysophosphatidic acid, a growth factor-like lipid, in the saliva. *J. Lipid Res.* **2002**, *43*, 2049–2055. [CrossRef] [PubMed]

37. Xu, K.-P.; Yin, J.; Yu, F.-S.X. Lysophosphatidic acid promoting corneal epithelial wound healing by transactivation of epidermal growth factor receptor. *Invest. Ophthalmol. Vis. Sci.* **2007**, *48*, 636–643. [CrossRef] [PubMed]

38. Tigyi, G. Physiological responses to lysophosphatidic acid and related glycero-phospholipids. *Prostaglandins Other Lipid Mediat.* **2001**, *64*, 47–62. [CrossRef]

39. Jalink, K.; Hordijk, P.L.; Moolenaar, W.H. Growth factor-like effects of lysophosphatidic acid, a novel lipid mediator. *BBA-Rev. Cancer* **1994**, *1198*, 185–196. [CrossRef]

40. Nishida, K. Tissue engineering of the cornea. *Cornea* **2003**, *22*, S28–S34. [CrossRef] [PubMed]

41. Shah, A.; Brugnano, J.; Sun, S.; Vase, A.; Orwin, E. The development of a tissue-engineered cornea: Biomaterials and culture methods. *Pediatr. Res.* **2008**, *63*, 535–544. [CrossRef] [PubMed]

42. Wang, H.-Y.; Wei, R.-H.; Zhao, S.-Z. Evaluation of corneal cell growth on tissue engineering materials as artificial cornea scaffolds. *Int. J. Ophthalmol.* **2013**, *6*, 873–878. [CrossRef] [PubMed]

43. Vázquez, N.; Rodríguez-Barrientos, C.A.; Aznar-Cervantes, S.D.; Chacón, M.; Cenis, J.L.; Riestra, A.C.; Sánchez-Avila, R.M.; Persinal, M.; Brea-Pastor, A.; Fernández-Vega Cueto, L.; et al. Silk fibroin films for corneal endothelial regeneration: Transplant in a rabbit descemet membrane endothelial keratoplasty. *Investig. Ophthalmol. Vis. Sci.* **2017**, *58*, 3357–3365. [CrossRef]

44. Madden, P.W.; Lai, J.N.X.; George, K.A.; Giovenco, T.; Harkin, D.G.; Chirila, T.V. Human corneal endothelial cell growth on a silk fibroin membrane. *Biomaterials* **2011**, *32*, 4076–4084. [CrossRef] [PubMed]

45. Melles, G.R.J.; Ong, T.S.; Ververs, B.; van der Wees, J. Descemet membrane endothelial keratoplasty (DMEK). *Cornea* **2006**, *25*, 987–990. [CrossRef] [PubMed]

46. Zhang, X.; Nakahara, Y.; Xuan, D.; Wu, D.; Zhao, F.-K.; Li, X.-Y.; Zhang, J.-S. Study on the optical property and biocompatibility of a tissue engineering cornea. *Int. J. Ophthalmol.* **2012**, *5*, 45–49. [CrossRef] [PubMed]

47. Joseph, N.; Prasad, T.; Raj, V.; Anil Kumar, P.R.; Sreenivasan, K.; Kumary, T.V. A cytocompatible poly(n-isopropylacrylamide-co-glycidylmethacrylate) coated surface as new substrate for corneal tissue engineering. *J. Bioact. Compat. Polym.* **2010**, *25*, 58–74. [CrossRef]

48. Kong, B.; Mi, S. Electrospun scaffolds for corneal tissue engineering: A review. *Materials (Basel)* **2016**, *9*. [CrossRef] [PubMed]

49. Ghezzi, C.E.; Rnjak-Kovacina, J.; Kaplan, D.L. Corneal Tissue Engineering: Recent Advances and Future Perspectives. *Tissue Eng. Part B Rev.* **2015**, *21*, 278–287. [CrossRef] [PubMed]

50. Joyce, N.C. Proliferative capacity of the corneal endothelium. *Prog. Retin. Eye Res.* **2003**, *22*, 359–389. [CrossRef]

51. Röder, A.; García-Gareta, E.; Theodoropoulos, C.; Ristovski, N.; Blackwood, K.; Woodruff, M. An Assessment of Cell Culture Plate Surface Chemistry for in Vitro Studies of Tissue Engineering Scaffolds. *J. Funct. Biomater.* **2015**, *6*, 1054–1063. [CrossRef] [PubMed]

52. Colombini, M. VDAC: the channel at the interface between mitochondria and the cytosol. *Mol. Cell. Biochem.* **2004**, *256–257*, 107–115. [CrossRef] [PubMed]

53. Wiener, M. Regeneration of the cornea. *J. Am. Med. Assoc.* **1909**, *LIII*, 762–765. [CrossRef]

54. Bernardo, A.A.; Bernardo, C.M.; Espiritu, D.J.; Arruda, J.A.L. The Sodium Bicarbonate Cotransporter: Structure, Function, and Regulation. *Semin. Nephrol.* **2006**, *26*, 352–360. [CrossRef] [PubMed]

nanomaterials

MDPI

Article

The Bioactivity and Photocatalytic Properties of Titania Nanotube Coatings Produced with the Use of the Low-Potential Anodization of Ti6Al4V Alloy Surface

Aleksandra Radtke [1,2,*], Adrian Topolski [1] ⓘ, Tomasz Jędrzejewski [3] ⓘ, Wiesław Kozak [3], Beata Sadowska [4], Marzena Więckowska-Szakiel [4], Magdalena Szubka [5], Ewa Talik [5], Lars Pleth Nielsen [6] and Piotr Piszczek [1,2]

[1] Faculty of Chemistry, Nicolaus Copernicus University in Toruń, Gagarina 7, 87-100 Toruń, Poland; topolski@umk.pl (A.T.); piszczek@chem.umk.pl (P.P.)
[2] Nano-Implant Ltd. Gagarina 5, 87-100 Toruń, Poland
[3] Faculty of Biology and Environmental Protection, Nicolaus Copernicus University in Toruń, Lwowska 1, 87-100 Torun, Poland; tomaszj@umk.pl (T.J.); wkozak@umk.pl (W.K.)
[4] Faculty of Biology and Environmental Protection, University of Łódź, Banacha 12/16, 90-237 Łódź, Poland; beata.sadowska@biol.uni.lodz.pl (B.S.); marzena.wieckowska@biol.uni.lodz.pl (M.W.-S.)
[5] A. Chełkowski Institute of Physics, University of Silesia, Uniwersytecka 4, 40-007 Katowice, Poland; magdalena.szubka@us.edu.pl (M.S.); Ewa.Talik@us.edu.pl (E.T.)
[6] Tribology Centre, Danish Technological Institute, Kongsvang Allé 29, 8000 Aarhus C, Denmark; lpn@teknologisk.dk
* Correspondence: aradtke@umk.pl; Tel.: +48-600-321-294

Received: 6 July 2017; Accepted: 21 July 2017; Published: 26 July 2017

Abstract: Titania nanotube (TNT) coatings were produced using low-potential anodic oxidation of Ti6Al4V substrates in the potential range 3–20 V. They were analysed by X-ray diffraction (XRD), Raman spectroscopy, X-ray photoelectron spectroscopy (XPS), and scanning electron microscopy (SEM). The wettability was estimated by measuring the contact angle when applying water droplets. The bioactivity of the TNT coatings was established on the basis of the biointegration assay (L929 murine fibroblasts adhesion and proliferation) and antibacterial tests against *Staphylococcus aureus* (ATCC 29213). The photocatalytic efficiency of the TNT films was studied by the degradation of methylene blue under UV irradiation. Among the studied coatings, the TiO_2 nanotubes obtained with the use of 5 V potential (TNT5) were found to be the most appropriate for medical applications. The TNT5 sample possessed antibiofilm properties without enriching it by additional antimicrobial agent. Furthermore, it was characterized by optimal biocompatibility, performing better than pure Ti6Al4V alloy. Moreover, the same sample was the most photocatalytically active and exhibited the potential for the sterilization of implants with the use of UV light and for other environmental applications.

Keywords: titania nanotubes; anodic oxidation; biointegration; antibacterial properties; photocatalytic activity

1. Introduction

The considerable progress within the field of biomaterials and their medical applications is a result of intensive development of materials science. There are numerous biomaterials that can be used in the human body, including metals, alloys, ceramics, synthetic, or natural polymers [1–7]. However, titanium (Ti) and titanium alloys are considered to be some of the most significant biomaterials due to their resistance towards body fluid effects, high corrosion resistance, great tensile strength, flexibility and biocompatibility. So far, they are the most widely used materials in implantology [8–12].

The responsibility for the biocompatibility of titanium and its alloys is attributed to the formation of a chemically stable and highly adherent thin protective passivation film of titanium oxide. The natural passivation oxide layer on titanium has a thickness about 3–8 nm and is formed spontaneously in the presence of air or oxidizing media. This is also the case in biological systems, where a bioliquid surrounds the metal [13–17]. In the case of implants, the stoichiometric defects and low stability of this film can lead to their delamination and loosening [18–20]. Therefore, it becomes necessary to make the biocompatible coating permanently bond with the surface of implants. The fabrication of titania coatings of a specified stoichiometry and morphology by the electrochemical anodization of Ti/Ti alloy surfaces is a way to achieve this aim [21–26]. This simple and inexpensive method can lead to the formation of titania nanotube coatings of desired and beneficial structure, morphology, dimensions (aspect ratio; hole diameter versus length of the nanotubes), as well as optimized physicochemical properties. According to previous reports, titania nanotube coatings are produced mainly in the anodic oxidation processes using 20 V or higher potential, and usually, they are annealed in order to obtain crystalline TiO_2 layers.

Taking into account the economic considerations and the noticeable general trend towards the use of energy-efficient and time-efficient processes, we have focused on titania nanotube coatings (TNT) in the present study, which were produced on the surface of Ti6Al4V alloy, using possibly low potentials, i.e., 3–20 V at short process time below 20 min and without subsequent annealing. On the other hand, we took into consideration the interaction between biomaterials and the microorganisms, since foreign body-associated infections (FBAIs) are still one of the most frequent and dangerous complications of modern implantology [27–29]. *Staphylococci* with their wide repertoire of surface adhesion and easy ability to form biofilm are among the microorganisms that most frequently may result in infections [30,31]. It was therefore apparent for the authors that modern implant systems should, if possible, not only actively participate in the integration with the bone of the recipient, but should also prevent microbial adhesion, biofilm formation and massive inflammation after the implantation.

To achieve this goal, it means to obtain the coatings, which possess the optimal ability to osseointegrate as well as antibacterial activity (without enriching them with additional agents, such as silver nanoparticles or antibiotics), the optimization of TNT fabrication processes has been carried out, and the results of the studies on above issues are the main part of this paper. Moreover, the studies on the correlation between the antimicrobial properties and the photocatalytic activity of TNT coatings have been included into the publication in order to present their potential application in UV sterilization of implants surface.

2. Results

The coatings consisting of vertically aligned titania nanotubes (TNT) were produced on the Ti6Al4V surface using the electrochemical anodization technique and known procedure [32]. Samples were produced in the potential range 3–20 V at room temperature, during a 20 min anodization process, in the presence of 0.3 wt. % aqueous hydrofluoric acid solution. Coatings obtained at the mentioned conditions were denoted as TNT3-TNT20. Analysis of TNT3-TNT20 SEM (Scanning Electron Microscopy) images revealed that uniform nanotube coatings of the same tube length (approximately 150–200 nm), open at the top and closed at the bottom, without cracks and gaps, were formed (Figure 1).

The results of BET (Brunauer–Emmett–Teller) investigations of TNT coatings produced on titanium alloy substrates showed that the value of the surface area of these coatings was decreasing as the applied anodization voltage was increasing, and were equal to 18.3, 16.8, 12.1, and 10.2 m^2/g for layers anodized at 4 V (TNT4), 6 V (TNT6), 15 V (TNT15), and 20 V (TNT20), respectively. The above findings were in good correlation with the data obtained for TNT produced on commercially pure titanium foils [33].

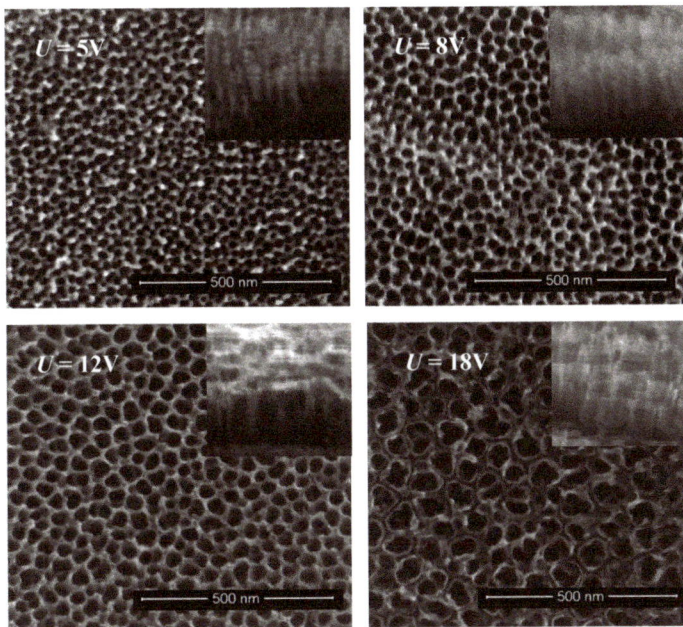

Figure 1. SEM (Scanning Electron Microscopy) images of the surface morphology and the cross section of titania nanotube (TNT) coatings on the surface of Ti6Al4V foil, produced at 5 V, 8 V, 12 V, and 18 V. Cross sections of the TNT coatings are illustrated as inserts.

2.1. Structural Characterization of TNT Coatings

Figure S1 in Supplementary Materials presents X-ray diffraction (XRD) patterns and Raman spectra of TNT coatings formed on the surface of titanium Ti6Al4V alloy. According to these data, materials produced on the surface of Ti6Al4V substrates, using the low potential anodic oxidation (3–20 V), were amorphous since no fingerprints of crystalline titanium dioxides could be seen neither in the XRD patterns nor in the Raman spectra.

In order to determine the nature of the oxide layer, the composition and the structure of produced TNT coatings were studied using X-ray photoelectron spectroscopy (XPS). The obtained data for as-received samples are presented in Table 1 and on Figure S2. Two peaks, which were found at binding energies of 459.0 eV and 464.8 eV, respectively, were attributed to titanium, $Ti(2p_{3/2})$ and $Ti(2p_{1/2})$ [34–36]. The splitting between the above-mentioned p-core levels is 5.8 eV (Table 1), which indicates the presence of a normal Ti^{4+} state in produced TNT coatings. The use of the deconvolution method revealed that the O(1s) peak can be composed of four (as it is visible for TNT4) or three (for TNT5-TNT18) components. The first component found at ~530.3 eV is attributed to O^{2-} in the Ti–O bond of TNT coatings. The second component located between 531.6 and 532.0 eV corresponds to oxygen of surface –OH groups. In this case, the splitting between the peaks is assigned to oxide species (TiO_2) and hydroxyl oxygen is 1.3–1.8 eV, and it is consistent with previous reports [34]. The components, which were found in the range 532.7–533.7 eV have been assigned to oxygen of water molecules adsorbed on the TNT oxidized surface (Table 1) [37,38]. The deconvolution of the O(1s) peak of TNT4 revealed the presence of a fourth component at 533.7 eV, which was attributed to the physically adsorbed water molecules on the surface of TNT layer [39,40]. Moreover, peaks, which were found at 285.0 eV (C–H/C–C), 286.4 eV (C–O), 289.0 eV (C=O), have been assigned to adsorbed carbon oxide and organic contaminants.

Table 1. X-ray photoelectron spectroscopy (XPS) data of selected TNT samples. The binding energies are in eV.

Sample	Ti^{4+}		O^{2-}	OH^{-}	H$_2$O	H$_2$O
	Ti(2p$_{3/2}$)	Ti(2p$_{1/2}$)	O(1s)	O(1s)	O(1s)	O(1s)
TNT4	459.0	464.8	530.3 (53%)	531.6 (19%)	532.7 (17%)	533.7 (11%)
TNT5	459.2	465.0	530.4 (67%)	531.8 (23%)	533.1 (10%)	
TNT6	459.0	464.8	530.3 (66%)	531.7 (24%)	533.0 (10%)	
TNT8	458.7	464.7	530.4 (64%)	531.9 (25%)	533.2 (11%)	
TNT10	458.8	464.6	530.2 (53%)	532.0 (32%)	533.4 (15%)	
TNT15	459.0	464.8	530.4 (69%)	531.8 (21%)	532.8 (10%)	
TNT18	458.7	464.5	530.3 (66%)	532.0 (23%)	533.4 (11%)	

2.2. The Wettability and the Roughness of TNT Coatings

The nature of hydrophobic and hydrophilic forces plays an important role both for the biological activity (impact on the cell adhesion and proliferation), as well as for the photocatalytic activity. Figure 2 shows the results of the wettability studies. They revealed the clear hydrophobic character of TNT5 and TNT6 samples. In both cases, the contact angle (θ) was close to 90 degrees and it was significantly higher in comparison to other TNT samples. The roughness of the produced coatings is another parameter influencing their biological and photocatalytic activity. Analysis of data presented in Figure 3 revealed that the roughness of TNT6-TNT10 layers is larger as compared to pure titanium alloy, and furthermore, larger than the roughness observed for TNT3-TNT5 and TNT12-TNT20 coatings.

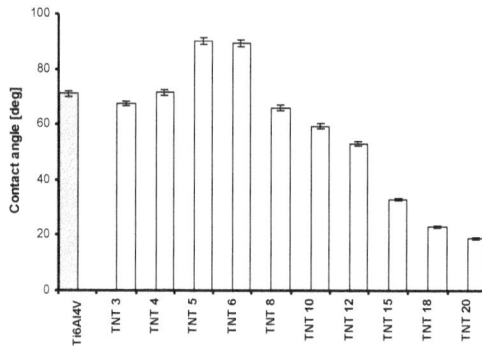

Figure 2. Results of wettability studies of TNT3-TNT20 samples.

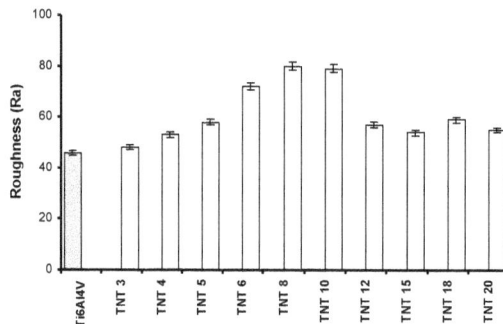

Figure 3. The surface roughness of TNT coatings determined by atomic force microscopy (AFM) data analysis.

2.3. Biological Activity of TNT Coatings Produced on the Surface of Ti6Al4V Foil

The biocompatibility of TNT coatings produced on the Ti6Al4V foil surface (Ti6Al4V-TNT system) was evaluated based on the MTT assay results, which were related to the adhesion (measured after 24 h) and proliferation (assessed after 72 h and 5 days) of L929 murine fibroblasts (Figure 4). It is worth noticing that all the studied samples showed higher level of fibroblasts proliferation than the reference samples after 72 h as well as after 5 days' incubation time. However, this effect was most noticeable in the case of TNT6-TNT10 samples, which consisted of densely packed nanotubes of ca. 25–35 nm in diameter.

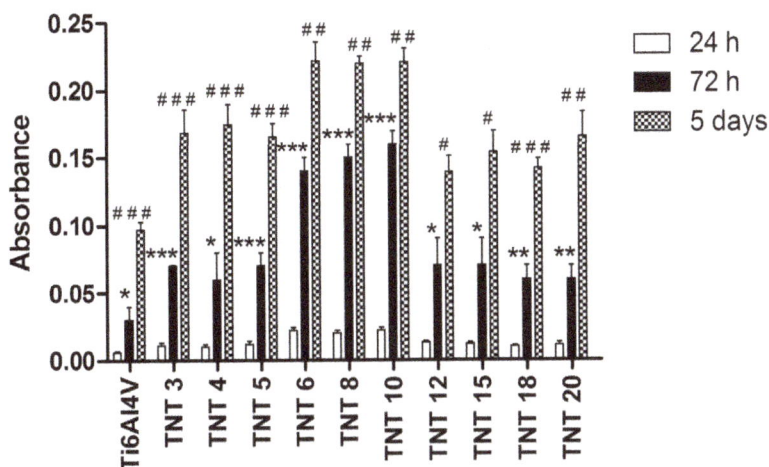

Figure 4. The effect of the incubation time on the murine L929 fibroblasts adhesion (after 24 h) and proliferation (after 72 h and 5 days) on the surfaces of Ti6Al4V alloy foils, modified by TNT, detected by MTT assays. The absorbance values are expressed as means ± S.E.M. of three experiments. Asterisk indicates significant differences between the cells incubated with the respective TNT for 24 h compared to 72 h of incubation time (* $P < 0.05$, ** $P < 0.01$, *** $P < 0.001$, respectively); hash mark denotes significant differences between the cells incubated with the same TNT for 72 h in comparison to 5 days' incubation time (# $P < 0.05$, ## $P < 0.01$, ### $P < 0.001$, respectively).

Figure 5 shows comparative micrographs of L929 murine fibroblasts cultured on the Ti6Al4V alloy and TiO$_2$ nanotubes: TNT5, TNT10, and TNT15 for 24 h (a, d, g, j), 72-h (b, e, h, k), and 5 days (c, f, i, l) respectively. Regarding the examination by SEM, the cells cultivated on the TNT surface effectively attached to the plate surface. Importantly, the fibroblast cultured on the plates for 24 h formed filopodia, which attached the cell to the surface of the plates, but did not form them among themselves (Figure 5m). This phenomenon was observed only after 72 h and 5 days of incubation time (Figure 5n,o, respectively). Moreover, particularly after 5 days, the fibroblasts incubated on plates were crowded and were forming networks due to overgrowth of cells, which indicates that the tested plates could contribute to the proliferation of the cells. The trend shown in the SEM analysis data in Figure 5 is the same as demonstrated in MTT assay (Figure 4). Furthermore, the cells have a more rounded shape after a 24-h incubation time, whereas those fibroblasts cultured for 72 h or 5 days on the TNT became increasingly more elongated and showed a number of filopodia.

The antibiofilm activity of Ti6Al4V-TNT system, produced with the use of anodic oxidation in the range of potentials 3–20 V was studied against *S. aureus* ATCC 29213 using colony-forming units (CFU) and LIVE/DEAD stained assays. The contact of the bacteria with the sample surfaces lasted 24 h. Figure 6 shows that inhibitory effect for *S. aureus* ATCC 29213 biofilm formations noticed in case of TNT coatings produced at 3–5 and 12 V. LIVE/DEAD-stained BacLight Bacterial Viability Kit confirmed clear antibiofilm activity of the TNT5 sample. The lowest values of green fluorescence units (Figure 7) and red fluorescence units (Figure 8), corresponding to the number of live and dead microorganisms, respectively, obtained for the coatings produced at 5 V.

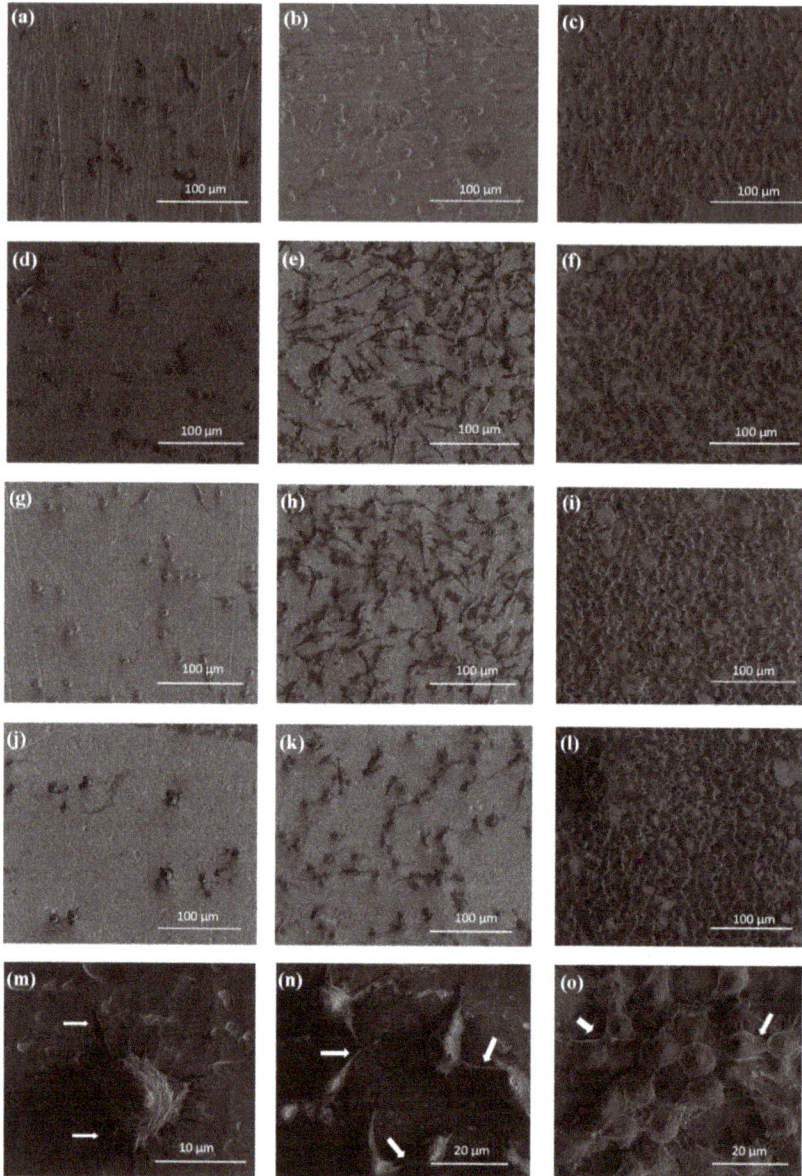

Figure 5. SEM images showing the cell adhesion (24 h) and proliferation (72 h and 5 days) of L929 murine fibroblasts on the Ti6Al4V alloy and TNT5, TNT10, and TNT15, after 24 h (**a,d,g,j**), 72 h (**b,e,h,k**), and 5 days (**c,f,i,l**) of incubation time, respectively. The arrows indicate the filopodia spread between fibroblasts incubated with TNT5 for 24 h, 72 h, and 5 days (**m–o**, respectively).

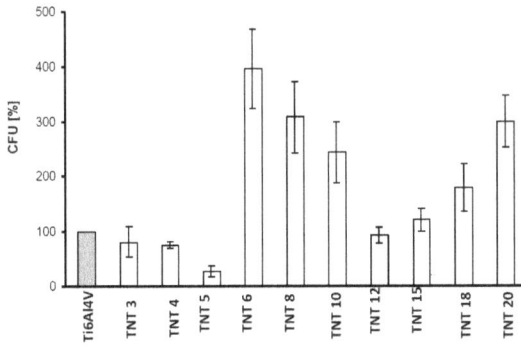

Figure 6. *S. aureus* ATCC 29213 aggregates/biofilm formation on the surfaces of Ti6Al4V alloy foil modified by TiO$_2$ nanotubes (TNT) tested by the colony-forming units (CFU) method. The results are presented as mean percentage ± standard deviation (S.D.) of *S. aureus* CFU reclaimed after 24 h from Ti6Al4V alloy biomaterials modified by TiO$_2$ nanotubes with different diameters (TNT), in comparison to bacterial CFU recovered from control (unmodified) biomaterial (Ti6Al4V) considered as 100%. Two independent sets of experiments were prepared, each in duplicate.

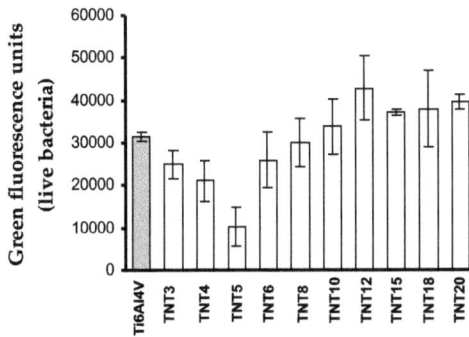

Figure 7. Number of live microorganisms adhered to Ti6Al4V alloy and TNT coatings, obtained in the anodic oxidation of Ti6Al4V surface.

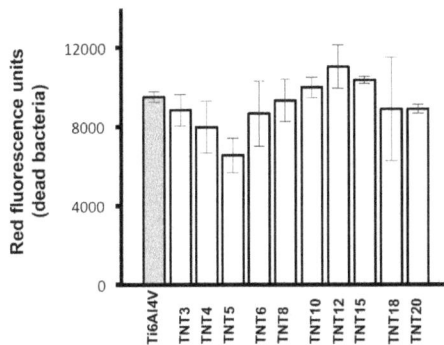

Figure 8. Number of dead microorganisms adhered to Ti6Al4V alloy and TNT coatings, obtained in the anodic oxidation of Ti6Al4V surface.

2.4. Photocatalytic Properties of TNT Coatings Produced on the Surface of Ti6Al4V Foil

The photodegradation of methylene blue (MB) is known in details [41]. This is why the MB photodegradation process is a good model for nanotubes photoactivity tests. The rate of MB photodegradation provides information on the photochemical activity of the studied TNT coatings produced at different anodization conditions. Additionally, it is a simple model system representative for organic water pollutant degradation. Figure S3 shows the absorbance of MB versus time dependence for the observed photodegradation process. All recorded kinetic measurements were similar in shape and could be fitted with one-exponential Equation (1),

$$A_t = A_{inf} - (A_{inf} - A_0) \exp(-k_{obs}t) \tag{1}$$

where A_t, A_{inf}, and A_0 represents absorbance in the real reaction time (t), infinity (inf), and start of the reaction ($t = 0$ s).

The observable rate constant is marked as k_{obs}, and values of this parameter, designated for samples TNT3-TNT20, have been summarized in Table 2. An analysis of these data exhibits that the values of k_{obs} change in the narrow range from 1.63 up to 1.95 ($\times 10^{-3}$ min^{-1}) and they do not seem to depend strongly on the morphology of the produced coatings and the obtained surface structure. However, the slightly higher photoactivity of samples TNT5, TNT10, TNT18 should be noted, which results in higher values of MB photodegradation rate constants (Table 2).

Table 2. The k_{obs} rate constants for methylene blue (MB) photodegradation on the TNT surfaces formed at different potentials. The results take into account blind tests (no UV and no titania samples).

Sample Rate Constant	TNT3	TNT4	TNT5	TNT6	TNT8	TNT10	TNT12	TNT15	TNT18	TNT20
$10^3 \times k_{obs}$ (min^{-1})	1.70 ± 0.14	1.63 ± 0.15	1.91 ± 0.13	1.62 ± 0.14	1.77 ± 0.16	1.89 ± 0.20	1.70 ± 0.16	1.80 ± 0.14	1.90 ± 0.16	1.69 ± 0.15

3. Discussion

The use of Ti6Al4V surface anodization allowed for the controlled formation of titania nanotube coatings of different tube diameters, i.e., from ca. 15 up to ca. 80 nm (Figure 9). Coatings produced between 3 V and 12 V consist of densely packed nanotubes, whereas the layers obtained at higher potentials (15–20 V) are formed by the separated nanotubes, as evident from the cross section images inserted in Figure 1. Nonlinear dependency between the used potential and the nanotubes diameter can be explained by the presence of the tubes separation processes, which proceed during the nanotubes growth process (Figure 9). An analysis of the SEM images revealed that the beginning of the nanotubes separation was observed for coatings obtained above 4 V, and the finish of the separation process was observed for samples produced at potentials higher than 12 V.

As the photo- and bioactivity of titania coatings is strongly dependent on their surface structure, it was decided to focus on the characterization of the amorphous surface structures in further details. According to McCafferty and Wightman, the metal-oxide system on the metal surface consists of three regions: (a) metal-oxide part, (b) hydroxylated part, and (c) chemisorbed water. This three-piece oxide layer is usually covered with adsorbed carbon oxides, organic contaminants from the air, as well as the physically adsorbed water [37]. The results of the XPS studies revealed that the percentage of adsorbed H_2O molecules and OH$^-$-groups on the surface of produced TNT coatings, changes depending on the condition of the anodic oxidation processes. TNT5 and TNT6, which consisted of the densely packed nanotubes of diameters of ca. 21–25 nm, are characterized by the low percentage of adsorbed H_2O molecules and the high concentration of OH$^-$-groups (Table 1). For the layer composed of nanotubes where the diameter was smaller than ca. 20 nm, (TNT4), the XPS studies confirmed the presence of chemisorbed, as well as physically adsorbed, water molecules and the low concentration of the hydroxyl groups. In turn, for nanotubes with diameters above ca. 25 nm, but still possessing the common walls, the amount of H_2O molecules and –OH groups increases. The effect of the nanotubes' final separation, which was noticed for TNT15, influences probably the insignificant amount of water molecules and hydroxyl groups on the surface of nanotubes.

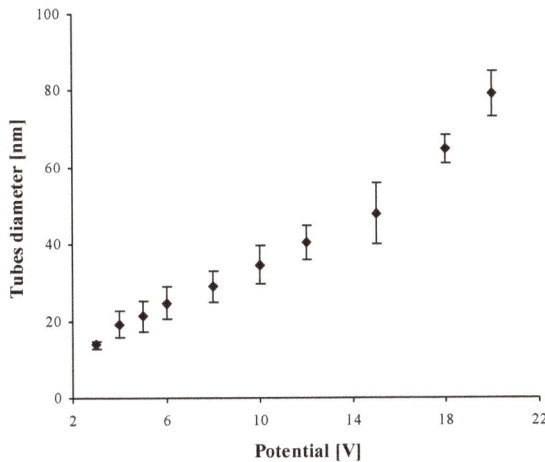

Figure 9. The diameter of titania nanotubes as a function of used potential.

According to earlier reports, titanium and its alloys revealed a more hydrophobic character. However, their anodic oxidation leads to the formation of more hydrophilic systems [42–46]. In the case of coatings composed of titania nanotubes, their hydrophilicity significantly depends on the nanotubes' diameter, e.g., coatings composed of large diameter nanotubes are more hydrophilic. It can be explained by the fact that capillary forces of the liquid are able to facilitate water penetration into the tube interior. The TNT coatings produced at 5 V and 6 V are composed of densely packed nanotubes of diameters ca. 21–25 nm. According to the XPS data, the surface of these materials characterizes relatively low adsorption of water molecules (Table 1), which is in accord with the contact angle findings. In the comparison to TNT5 and TNT6, an increase in the hydrophilicity has been observed for both TNT surfaces having nanotubes of smaller diameters (below ca. 21 nm), as well as for TNT surfaces having larger nanotube diameters (above ca. 25 nm). The further increase in the hydrophilicity of the produced coatings is associated with the increase of the nanotube diameter, according to the trend seen by other authors in the literature [47]. The rapid increase in the TNT coatings hydrophilicity for materials produced between 12 and 20 V is associated with the separation process of nanotubes, which proceeds on the substrate surface (Figures 1 and 9). Considering the obtained results, it should be noted that the nanotube diameters and their separation are the main factors influencing the wettability properties of the studied TNT coatings.

We have previously shown that the adhesion and the proliferation of fibroblasts on the surface of Ti-TNT system were significantly higher than on the surface of pure non-oxidized titanium [32]. According to literature reports, the use of Ti6Al4V alloy as a substrate offers much better physical and mechanical properties than pure titanium, as well as excellent biocompatibility [48]. The results of the studies of Ti6Al4V-TNT system, produced by the use of various potentials (3–20 V), revealed that the adhesion and the proliferation of L929 cells were greater as compared to the nonoxidized reference sample (pure Ti6Al4V) (Figure 4). The smaller differences in the cell proliferation after 5 days ($P < 0.01$ for 72 h versus $P < 0.05$ for 5 days) may be due to the fact that the fibroblasts were crowded and formed network due to overgrowth of cells. Furthermore, the cells overgrowing the entire surface of the plates did not have enough free space for further subdivisions. This assumption was confirmed by the results of the SEM analysis (Figure 5).

In general, the results of the MTT assay confirmed the promising biocompatible properties of all produced TNT coatings. They give hope for the use of studied coatings as biomaterials in implantology, since favorable cellular interaction with their surface is crucial to the long-term success of implants [49]. Analysis of data presented in Figure 4 revealed the lack of significant differences in terms of the adhesion and the proliferation of cells on the surface of TNT3-TNT5 and also TNT12-TNT20 samples. The biointegration was most noticeable in the TNT6-TNT10 samples. The mentioned greater fibroblast cells adhesion and proliferation may be associated with the high roughness values of TNT6-TNT10 (Figures 3 and 4).

The inhibitory effect for *S. aureus* ATCC 29213 biofilm formations was noticed in the TNT coatings produced at 3–5 and 12 V. It was intensified with the increase of the surface hydrophobicity and it was the strongest for TNT5. This effect was observed to be weaker for TNT12-TNT20 samples, what could be associated with the increase of the hydrophilicity for these coatings consequentially with the increase of the nanotubular diameter and their separation (Figures 2 and 9). The stimulation effect of the biofilm formation, which was noted for TNT6-TNT10 samples is incomprehensible and requires further explanation. The enhanced adhesion of bacteria to the nanotubular and nanotextured surfaces, is speculated by Puckett as being a result of their amorphousness and the greater nanometer surface roughness [50]. Our studies have shown that all the produced TNT coatings were amorphous, which together with the high surface roughness of TNT6-TNT10 may lead to a significant increase in their vulnerability on bacterial attachment in comparison to the conventional Ti6Al4V non-anodized surfaces. Puckett et al. also pointed out the ambiguous role of fluoride ions, which are present on the nanotubular titanium surfaces [50]. According to them, the fluorine present on the TNT coating surface may increase the adhesion of bacteria. On the other hand, the earlier studies confirmed the antibacterial effect caused by the presence of fluorine [51–53]. Results of our XPS studies confirmed that TNT coatings formed during the anodization processes contain fluorine ions, and furthermore, that the fluorine content is different for TNT samples obtained at different potentials (Table 3). An analysis of the XPS data indicated that the highest fluorine concentration is observed on the TNT5 surface, which might be linked with the good antimicrobial properties observed for this coating.

Table 3. Fluorine ions' presence on the surface of TNT coatings based of XPS studies.

Sample	TNT4	TNT5	TNT6	TNT8	TNT10	TNT15	TNT18
F %	4.2	6.3	4.5	1.5	2.9	2.4	1.3

The LIVE/DEAD assay also confirmed clear antibiofilm activity of TNT5, as the lowest number of live and dead bacteria (Figures 7 and 8) was noticed for this coating. This indicates that TNT5 is the most active surface preventing bacterial adhesion and biofilm formation. Moreover, based on the suggestions of Puckett et al., and Palma et al. it could be suspected that the growing number of dead bacteria on the selected titanium surfaces (TNT6-TNT15), which can release an intercellular protein upon death, might become the nutrient for others and enhance further adhesion to other microorganisms [50,54]. In accordance with this, the number of live staphylococci on those surfaces is also observed to increase.

Making the discussion about the photocatalytic activity of obtained nanotube coatings we can state that, generally, the values of k_{obs} for TNT coatings produced on the surface of Ti6Al4V are lower in comparison to previously noticed data for TNT on Ti substrates [33]. The observed difference can be explained by the amorphousness of the present TNT as well as the presence of smaller diameters of nanotubes on alloy surfaces, in comparison to TNT layers formed on titanium substrates. Nonetheless, although TNT formed on the titanium alloy reacts slower, the difference which is not distinctly significant (the same order of magnitude) and popularity of this alloy in many applications (e.g., medical) makes TNT formed on Ti6Al4V a very interesting material, which can compete with TNT formed on the pure titanium foil, if photochemical properties are taken into account.

Among the studied Ti6Al4V-TNT systems, obtained with the use of low-potential anodic oxidation of titanium alloy, only the TNT5 surface appears to maintain an appropriate balance between the tissue biocompatibility allowing for the colonization of the host eukaryotic cells and the ability to prevent the bacterial adhesion and biofilm formation. Therefore, we can conclude that TNT5 coating is superior for biomedical applications as implants surface coating. Moreover, TNT5 is the most photoactive among the studied materials, in the degradation of MB solution, so it can be used simultaneously as active coating in the process of implant surface sterilization induced by UV light.

Biological properties of TNT6, TNT8, and TNT10, which are characterized by the highest roughness compared to other produced TNT coatings, are noteworthy. These materials revealed the best properties for the adhesion and proliferation of fibroblasts. However, at the same time they are also characterized by a surprisingly high adhesion of microorganisms and tendency to biofilm formation, which excludes their use as biomedical coatings.

It should be pointed out that the obtained results have a limitation—they are adequate for Ti6Al4V alloy, which is commonly used in maxillofacial and dental implantology. In the case of the other titanium alloy use (for example Ti6Al7Nb and Ti13Nb13Zr used in orthopaedics), the optimization process of TNT production is necessary.

4. Materials and Methods

4.1. Synthesis of TiO$_2$ Nanotubes Coatings (TNT)

TiO$_2$ nanotubes (TNT) were produced on the surface of Ti6Al4V alloy samples (5 mm × 70 mm, Grade 5, BIBUS METALS) using the anodic oxidation method. Before the process of anodization, the substrate samples were ultrasonically cleaned sequentially in acetone (15 min), 80% ethanol (5 min), and deionised water (15 min). The substrates were dried in an Argon stream at room temperature. The surface of the substrates were chemically etched in a 1:4:5 mixture of HF:HNO$_3$:H$_2$O for 30 s, cleaned with deionised water, and dried in an argon (Ar) stream. The anodization was carried out at room temperature using prepared substrate as anode, platinum wire as cathode, and 0.3 wt. % aqueous HF solution as electrolyte, according to earlier reports [32]. The applied potential was varied from 3 V up to 20 V and the anodization time $t = 20$ min. In order to purify the produced coatings, they were washed with distilled water with the addition of Al$_2$O$_3$ powder (averaged particle size = 50 nm) in an ultrasonic bath for 1 min, and then dried in Ar stream. Samples obtained at mentioned conditions were denoted as: TNT3-TNT20.

4.2. Morphological and Structural Characterization of TNT Coatings

The morphology of the produced coatings was studied using Quanta field-emission gun Scanning Electron Microscope (SEM; Quanta 3D FEG; Carl Zeiss, Göttingen, Germany; 30.0 kV accelerating voltage was chosen for SEM analysis and the micrographs were recorded under high vacuum using secondary electron detector (SE)). The surface roughness of the produced coatings was established based on atomic force microscopy studies (AFM; NanoScope MultiMode SPM System, Bruker, Billerica, MA, USA, with scanning probe Veeco Digital Instrument, measurement in the tapping mode (noncontact mode), scan area: 5 × 5 μm). The structure of the produced TiO$_2$ nanotube layers (TNT) was analyzed using X-ray diffraction (PANalytical X'Pert Pro MPD X-ray diffractometer, PANalytical B.V., Almelo, The Netherlands, using Cu-Kα radiation; the incidence angle was equal to 1 deg) and Raman spectroscopy (RamanMicro 200, Perkin Elmer, Waltham, MA, USA, Exposure time 2 s; Number of exposure 20; Spectral range 200–3200 cm^{-1}; number of scanned points on the sample surface −10). XPS spectra of investigated samples were obtained with monochromatized Al K$_\alpha$-radiation (1486.6 eV) at room temperature using a PHI 5700/660 ESCA spectrometer (Physical Electronics, Lake Drive East Chanhassen, MN, USA). Studies on BET-specific surface area were done using the Accelerated Surface Area and Porosimetry System ASAP 2010 (Micromeritics, Norcross, GA, USA). The samples were heated (desorbed) before measurement at 70 °C to achieve a final pressure of 0.001 mbar, over 8 h.

After the desorption process, the samples were weighed and placed in a measuring station in the temperature of liquid nitrogen, in which the nitrogen adsorption isotherms were determined).

4.3. Wettability Measurements

The wettability of TNT coatings was investigated using a Drop Shape Analyzer—DSA100S (Krüss, Hamburg, Germany) for the contact angle measurement. Ten µL of distilled water were slowly deposited on the surface of analyzed TNT coatings using a calibrated screw-syringe. The images were recorded and the contact angles were estimated by numerically fitting of the droplet images. The value of the contact angle for each biomaterial is the average value of five measurements.

4.4. Cell Adhesion and Proliferation Assay on TNT Coatings

Murine fibroblasts cell line L929 (American Type Culture Collection) culture conditions were the same as described previously [32]. The effect of TNT on the cells adhesion (after 24 h) and proliferation (after 72 h and 5 days, respectively) were studied by the MTT (3-(4,5-dimethylthiazole-2-yl)-2,5-diphenyl tetrazolium bromide; Sigma Aldrich, Darmstadt, Germany) assay using the same method as it was reported in [32]. The morphology changes of L929 cells grown on the surface of TiO_2 nanotubes coatings were analyzed using Scanning Electron Microscopy (SEM, Quanta 3D FEG; Carl Zeiss, Göttingen, Germany).

4.5. Microbial Aggregates/Biofilm Formation on TNT Coatings

TNT coatings on the surface of Ti6Al4V alloy substrates, prepared in the accordance with the procedure previously used [32], have been exposed to *Staphylococcus aureus* ATCC 29213 reference strain. The samples (size ~5 mm × 5 mm) of studied TNT layers and unmodified Ti6Al4V alloy (control sample) were placed into *S. aureus* suspension (OD = 0.9) for 24 h and incubated in stable conditions at 37 °C to check the formation of microbial aggregates/biofilm on the surface of TNT coatings [32]. To evaluate aggregates/biofilm formation, a CFU method was used after mechanical recovery of microbial cells from the tested surfaces, as well as LIVE/DEAD-stained BacLight Bacterial Viability kit (L/D; Invitrogen, Thermo Fisher Scientific, Eugene, OA, USA).

4.6. Photocatalytic Degradation of Methylene Blue (MB)

Studies on the photocatalytic degradation of methylene blue (MB) were performed using MB aqueous solution of initial concentration $c_0 = 1.0 \times 10^{-5}$ M, according to the procedure written in earlier reports [33]. The kinetic calculations are based on the methodology of chemical kinetics assuming a Langmuir–Hinshelwood reaction mechanism. Taking into account a low concentration of MB, it can be assumed that a photodegradation process occurs according to the pseudo-first-order kinetics, and the kinetic equation describing changes in the MB concentration during its degradation can be expressed as below (Equation (2)):

$$c_t = c_0 \exp(-k_{obs}t), \tag{2}$$

where c_t is MB concentration after time t, c_0 is its starting concentration, and k_{obs} is the observable rate constant. In the calculations, blind tests (degradation of MB with no UV and no titania samples) were taken into account.

Supplementary Materials: The following are available online at http://www.mdpi.com/2079-4991/7/8/197/s1, Figure S1: Results of XRD studies of TNT coatings produced at 5, 8, 12, and 18V, respectively (a) and Raman spectra of these materials (b); Figure S2: XPS spectra of selected TNT samples, O1s peaks after deconvolution process.; Figure S3: Changes of MB absorbance as a function of time with UV light illumination in the presence of TNT3-TNT20.

Acknowledgments: The studies were supported by MP1301 COST Action NEWGEN (Short-Term Scientific Mission grant awarded to A. Radtke; COST-STSM-ECOST-MP1301-010614-045087). Technology Transfer Centre of Nicolaus Copernicus University in Torun is acknowledged for covering the costs of publishing in open access.

Author Contributions: Aleksandra Radtke, Piotr Piszczek and Lars Pleth Nielsen conceived and designed the experiments; Aleksandra Radtke performed the chemical experiments and the characterization of obtained materials; Adrian Topolski analyzed the photocatalytic data; Tomasz Jędrzejewski performed the immunological experiments; Wiesław Kozak analyzed the biointegration studies data and provided necessary equipment and materials to immunological experiments; Beata Sadowska, Marzena Więckowska-Szakiel performed the microbiological experiments; Magdalena Szubka performed XPS analysis; Ewa Talik provided XPS analysis; Aleksandra Radtke, Tomasz Jędrzejewski, Beata Sadowska, and Piotr Piszczek analyzed the data; Aleksandra Radtke wrote the paper.

Conflicts of Interest: The authors declare no conflicts of interest.

References

1. Williams, D.F. *The Williams Dictionary of Biomaterials*; Liverpool University Press: Liverpool, UK, 2008; Volume 42.
2. Sáenz, A.; Rivera-Muñoz, E.; Brostow, W.; Castaño, V.M. Ceramic biomaterials: An introductory overview. *J. Mater. Educ.* **1999**, *21*, 297–306.
3. Kokubo, T.; Kim, H.M.; Kawashita, M. Novel bioactive materials with different mechanical properties. *Biomaterials* **2003**, *13*, 2161–2175. [CrossRef]
4. Kokubo, T.; Kim, H.M.; Kawashita, M.; Nakamura, T. Bioactive metals: Preparation and properties. *J. Mater. Sci. Mater. Med.* **2004**, *15*, 99–107. [CrossRef] [PubMed]
5. Park, J.P.; Bronzino, J.D. Biomaterials: Principles and Applications. In *Metallic Biomaterials*; Kim, Y.K., Park, J.B., Eds.; CRC Press LLC: Boca Raton, FL, USA, 2003; pp. 1–20.
6. Teo, A.J.T.; Mishra, A.; Park, I.; Kim, Y.-J.; Park, W.-T.; Yoon, Y.-J. Polymeric biomaterials for medical implants and devices. *ACS Biomater. Sci. Eng.* **2016**, *2*, 454–472. [CrossRef]
7. Ananth, H.; Kundapur, V.; Mohammed, H.S.; Anand, M.; Amarnath, G.S.; Mankar, S. Review on biomaterials in dental implantology. *Int. J. Biomed. Sci.* **2015**, *11*, 113–120. [PubMed]
8. Hench, L.L.; Polak, J.M. Third-Generation Biomedical Materials. *Science* **2002**, *295*, 1014–1017. [CrossRef] [PubMed]
9. Manivasagam, G.; Dhinasekaran, D.; Rajamanickam, A. Biomedical Implants: Corrosion and its Prevention—A Review. *Recent Pat. Corros. Sci.* **2010**, *2*, 40–54. [CrossRef]
10. Özcan, M.; Hämmerle, C.H. Titanium as a Reconstruction and Implant Material in Dentistry: Advantages and Pitfalls. *Materials* **2012**, *5*, 1528–1545.
11. Geetha, M.; Singh, A.K.; Asokamani, R.; Gogia, A.K. Ti based biomaterials, the ultimate choice for orthopaedic implants—A review. *Prog. Mater. Sci.* **2009**, *54*, 397–425. [CrossRef]
12. Ratner, B. *Titanium in Medicine: Material Science, Surface Science, Engineering, Biological Responses and Medical Applications*; Brunette, D.M., Tengvall, P., Textor, M., Thompson, P., Eds.; Springer: Berlin/Heidelberg, Germany, 2001; pp. 1–12.
13. Cui, C.; Liu, H.; Li, Y.; Sun, J.; Wang, R.; Liu, S.; Greer, A.L. Fabrication and biocompatibility of nano-TiO_2/titanium alloys biomaterials. *Mater. Lett.* **2005**, *59*, 3144–3148. [CrossRef]
14. Al-Mobarak, N.A.; Al-Swayih, A.A. Development of Titanium Surgery Implants for Improving Osseointegration Through Formation of a Titanium Nanotube Layer. *Int. J. Electrochem. Sci.* **2014**, *9*, 32–45.
15. López-Huerta, F.; Cervantes, B.; González, O.; Hernández-Torres, J.; García-González, L.; Vega, R.; Herrera-May, A.L.; Soto, E. Biocompatibility and Surface Properties of TiO_2 Thin Films Deposited by DC Magnetron Sputtering. *Materials* **2014**, *7*, 4105–4117. [CrossRef]
16. Majeed, A.; He, J.; Jiao, L.; Zhong, X.; Sheng, Z. Surface properties and biocompatibility of nanostructured TiO_2 film deposited by RF magnetron sputtering. *Nanoscale Res. Lett.* **2015**, *10*, 56. [CrossRef] [PubMed]
17. Bianchi, M. *Multiscale Fabrication of Functional Materials for Regenerative Medicine*; Springer: Berlin/Heidelberg, Germany, 2011; p. 45.
18. Heimann, R.B. Thermal spraying of biomaterials. *Surf. Coat. Technol.* **2006**, *201*, 2012–2019. [CrossRef]
19. Oshida, Y.; Tuna, E.B.; Aktören, O.; Gençay, K. Dental Implant Systems. *Int. J. Mol. Sci.* **2010**, *11*, 1580–1678. [CrossRef] [PubMed]
20. Sobieszczyk, S.; Klotzke, R. Nanotubular titanium oxide layers for enhancement of bone-implant bonding and bioactivity. *Adv. Mater. Sci.* **2011**, *11*, 17–26. [CrossRef]

21. Roy, P.; Berger, S.; Schmuki, P. TiO$_2$ Nanotubes: Synthesis and Applications. *Angew. Chem. Int. Ed.* **2011**, *50*, 2904–2939. [CrossRef] [PubMed]
22. Minagar, S.; Berndt, C.C.; Wang, J.; Ivanova, E.; Wen, C. A review of the application of anodization for the fabrication of nanotubes on metal implant surfaces. *Acta Biomater.* **2012**, *8*, 2875–2888. [CrossRef] [PubMed]
23. Macak, J.M.; Tsuchiya, H.; Ghicov, A.; Yasuda, K.; Hahn, R.; Bauer, S.; Schmuki, P. TiO$_2$ nanotubes: Self-organized electrochemical formation, properties and applications. *Curr. Opin. Solid State Mater. Sci.* **2007**, *11*, 3–18. [CrossRef]
24. Macak, J.M.; Hildebrand, H.; Marten-Jahns, U.; Schmuki, P. Mechanistic aspects and growth of large diameter self organized TiO$_2$ nanotubes. *J. Electroanal. Chem.* **2008**, *621*, 254–266. [CrossRef]
25. Tan, A.W.; Pingguan-Murphy, B.; Ahmad, R.; Akbar, S.A. Review of titania nanotubes: Fabrication and cellular response. *Ceram. Int.* **2012**, *38*, 4421–4435. [CrossRef]
26. Brammer, K.S.; Frandsen, C.J.; Jin, S. TiO$_2$ nanotubes for bone regeneration. *Trends Biotechnol.* **2012**, *30*, 315–322. [CrossRef] [PubMed]
27. Arciola, C.R.; Montaro, L.; Costerton, J.W. New trends in diagnosis and control strategies for implant infections. *Int. J. Artif. Organs* **2011**, *34*, 727–736. [CrossRef] [PubMed]
28. Montanaro, L.; Speziale, P.; Campoccia, D.; Ravaioli, S.; Cangini, I.; Pietrocola, G.; Giannini, S.; Arciola, C.R. Scenery of *Staphylococcus* implant infections in orthopedics. *Future Microbiol.* **2011**, *6*, 1329–1349. [CrossRef] [PubMed]
29. Yue, C.; Yang, B. Bioactive titanium surfaces with the effect of inhibiting biofilm formation. *J. Bionic Eng.* **2014**, *11*, 589–599. [CrossRef]
30. Hanke, M.L.; Kielian, T. Deciphering mechanisms of staphylococcal biofilm evasion of host immunity. *Front. Cell. Infect. Microbiol.* **2012**, *2*, 62. [CrossRef] [PubMed]
31. Zhao, G.; Zhong, H.; Zhang, M.; Hong, Y. Effects of antimicrobial peptides on *Staphylococcus aureus* growth and biofilm formation in vitro following isolation from implant-associated infections. *Int. J. Clin. Exp. Med.* **2015**, *8*, 1546–1551. [PubMed]
32. Lewandowska, Ż.; Piszczek, P.; Radtke, A.; Jędrzejewski, T.; Kozak, W.; Sadowska, B. The evaluation of the impact of titania nanotube covers morphology and crystal phase on their biological properties. *J. Mater. Sci. Mater. Med.* **2015**, *26*, 163. [CrossRef] [PubMed]
33. Radtke, A.; Piszczek, P.; Topolski, A.; Hald Andersen, I.; Pleth Nielsen, L.; Heikkilä, M.; Leskelä, M. The evaluation of the impact of titania nanotube covers morphology and crystal chase on their biological properties. *Appl. Surf. Sci.* **2016**, *368*, 165–172. [CrossRef]
34. Lai, Y.; Sun, L.; Chen, Y.; Zhuang, H.; Lin, C.; Chin, J.W. Effects of the structure of TiO$_2$ nanotube array on Ti substrate on its photocatalytic activity. *J. Electrochem. Soc.* **2006**, *153*, 123–127. [CrossRef]
35. Akhavan, O.; Ghaderi, E. Capping antibacterial Ag nanorods aligned on Ti interlayer by mesoporous TiO$_2$ layer. *Surf. Interface Anal.* **2009**, *203*, 3123–3128. [CrossRef]
36. Akhavan, O.; Ghaderi, E. Self-accumulated Ag nanoparticles on mesoporous TiO$_2$ thin film with high bactericidal activities. *Surf. Coat. Technol.* **2010**, *204*, 3676–3683. [CrossRef]
37. McCafferty, E.; Wightman, J.P. Determination of the concentration of surface hydroxyl groups on metal oxide films by a quantitative XPS method. *Surf. Interface Anal.* **1998**, *26*, 549–564. [CrossRef]
38. Tan, X.; Fan, Q.; Wang, X.; Grambow, B. Eu(III) sorption to TiO$_2$ (anatase and rutile): Batch, XPS, and EXAFS studies. *Environ. Sci. Technol.* **2009**, *43*, 3115–3121. [CrossRef] [PubMed]
39. Haija, M.A.; Guimond, S.; Uhl, A.; Kuhlenbeck, H.; Freund, H.-J. Adsorption of water on thin V$_2$O$_3$(0001) films. *Surf. Sci.* **2006**, *600*, 1040–1047. [CrossRef]
40. Henderson, M.A. Water adsorption on TiO$_2$ surfaces probed by soft X-ray spectroscopies: Bulk materials vs. isolated nanoparticles. *Surf. Sci. Rep.* **2002**, *46*, 15088.
41. Houas, A.; Lachheb, H.; Ksibi, M.; Elaloui, E.; Guillard, C.; Herrmann, J.-M. Photochemical Characterization and Photocatalytic Properties of a Nanostructure Composite TiO$_2$ Film. *Appl. Catal. B Environ.* **2001**, *31*, 145–157. [CrossRef]
42. Ninham, B.W.; Kurihara, K.; Vinogradova, O. Hydrophobicity, specific ion adsorption and reactivity. *Colloid Surf. A* **1997**, *123–124*, 7–12. [CrossRef]
43. Park, J.; Bauer, S.; Schleger, K.A.; Neukam, F.W.; Von Mark, K.D.; Schmuki, P. TiO$_2$ nanotube surfaces: 15 nm–an optimal length scale of surface topography for cell adhesion and differentiation. *Small* **2009**, *5*, 666–671. [CrossRef] [PubMed]

44. Brammer, K.S.; Oh, S.; Gallagher, J.O.; Jin, S. Enhanced cellular mobility guided by TiO$_2$ nanotube surfaces. *Nano Lett.* **2008**, *8*, 786–793. [CrossRef] [PubMed]
45. Sun, T.; Wang, M. A comparative study on titania layers formed on Ti, Ti-6Al-4V and NiTi shape memory alloy through a low temperature oxidation process. *Surf. Coat. Technol.* **2010**, *205*, 92–101. [CrossRef]
46. Clem, W.C.; Konovalov, V.; Chowdhury, S.; Vohra, Y.K.; Catledge, S.A.; Bellis, S.L. Mesenchymal stem cell adhesion and spreading on microwave plasma-nitrided titanium alloy. *J. Biomed. Mater. Res. A* **2006**, *76*, 279–287. [CrossRef] [PubMed]
47. Shin, D.H.; Shokuhfar, T.; Choi, C.K.; Lee, S.-H.; Friedrich, C. Wettability changes of TiO$_2$ nanotube surfaces. *Nanotechnology* **2011**, *22*, 315704. [CrossRef] [PubMed]
48. Giordano, C.; Saino, E.; Rimondini, L.; Pedeferri, M.P.; Visai, L.; Cigada, A.; Chiesa, R. Electrochemically induced anatase inhibits bacterial colonization on Titanium Grade 2 and Ti6Al4V alloy for dental and orthopedic devices. *Colloids Surf. B Biointerfaces* **2011**, *88*, 648–655. [CrossRef] [PubMed]
49. Cevc, G.; Vierl, U. Nanotechnology and the transdermal route: A state of the art review and critical appraisal. *J. Control. Release* **2010**, *141*, 277–299. [CrossRef] [PubMed]
50. Puckett, S.D.; Taylor, E.; Raimondo, T.; Webster, T.J. The relationship between the nanostructure of titanium surfaces and bacterial attachment. *Biomaterials* **2010**, *31*, 706–713. [CrossRef] [PubMed]
51. Yoshinari, M.; Oda, Y.; Kato, T.; Okuda, K. Influence of surface modifications to titanium on antibacterial activity in vitro. *Biomaterials* **2001**, *22*, 2043–2048. [CrossRef]
52. Raulio, M.; Jarn, M.; Ahola, J.; Peltonen, J.; Rosenholm, J.B.; Tervakangas, S. Microbe repelling coated stainless steel analysed by field emission scanning electron microscopy and physicochemical methods. *J. Ind. Microbiol. Biotechnol.* **2008**, *35*, 751–760. [CrossRef] [PubMed]
53. Hyde, F.W.; Alberg, M.; Smith, K. Comparison of fluorinated polymers against stainless steel, glass and polypropylene in microbial biofilm adherence and removal. *J. Ind. Microbiol. Biotechnol.* **1997**, *19*, 142–149. [CrossRef] [PubMed]
54. Palma, M.; Haggar, A.; Flock, J.I. Adherence of *Staphylococcus aureus* is enhanced by an endogenous secreted protein with broad binding activity. *J. Bacteriol.* **1999**, *181*, 2840–2845. [PubMed]

nanomaterials

MDPI

Article

Gadolinium Tagged Osteoprotegerin-Mimicking Peptide: A Novel Magnetic Resonance Imaging Biospecific Contrast Agent for the Inhibition of Osteoclastogenesis and Osteoclast Activity

Lubinda Mbundi [1,2], Steve T. Meikle [2], Rosa Busquets [3], Nicholas G. Dowell [4], Mara Cercignani [4] and Matteo Santin [2,*]

[1]   Department of Surgical Research, Northwick Park Institute for Medical Research, University College London (UCL), Northwick Park & St Marks Hospitals, Watford Road, Harrow, Middlesex HA1 3UJ, UK; l.mbundi@ucl.ac.uk

[2]   Centre for Regenerative Medicine and Devices, School of Pharmacy and Biomolecular Sciences, University of Brighton, Huxley Building, Lewes Road, Brighton BN2 4GJ, UK; stevenmeikle@googlemail.com

[3]   Faculty of Science, Engineering and Computing, Penrhyn Road, Kingston University, Kingston Upon Thames KT1 2EE, UK; r.busquets@kingston.ac.uk

[4]   Clinical Imaging and Science Centre (CISC), Centre for Regenerative Medicine and Devices, Brighton and Sussex Medical School, Lewes Road, Brighton BN1 9RR, UK; N.G.Dowell@bsms.ac.uk (N.G.D.); M.Cercignani@bsms.ac.uk (M.C.)

*   Correspondence: m.santin@brighton.ac.uk; Tel.: +44-(0)-1273-642083

Received: 17 April 2018; Accepted: 28 May 2018; Published: 2 June 2018

check for updates

Abstract: The control of osteoblast/osteoclast cross-talk is crucial in the bone remodelling process and provides a target mechanism in the development of drugs for bone metabolic diseases. Osteoprotegerin is a key molecule in this biosignalling pathway as it inhibits osteoclastogenesis and osteoclast activation to prevent run-away bone resorption. This work reports the synthesis of a known osteoprotegerin peptide analogue, YCEIEFCYLIR (OP3-4), and its tagging with a gadolinium chelate, a standard contrast agent for magnetic resonance imaging. The resulting contrast agent allows the simultaneous imaging and treatment of metabolic bone diseases. The gadolinium-tagged peptide was successfully synthesised, showing unaltered magnetic resonance imaging contrast agent properties, a lack of cytotoxicity, and dose-dependent inhibition of osteoclastogenesis in vitro. These findings pave the way toward the development of biospecific and bioactive contrast agents for the early diagnosis, treatment, and follow up of metabolic bone diseases such as osteoporosis and osteosarcoma.

Keywords: osteoclastogenesis; RANK-RANKL-OPG; mimetic peptide; Gadolinium chelate; MRI

---

1. Introduction

The physiological remodelling of bone tissue is controlled by a delicate balance between the activity of the cells producing its mineralised extracellular matrix, the osteoblasts, and that of the osteoclasts, the cells responsible for its resorption [1,2]. This process of tissue turnover is regulated through the interaction of the receptor activator of nuclear factor kappa-B (RANK), expressed by osteoclast progenitors, with its cognate ligand (RANKL), which is expressed by the osteoblasts either as a membrane-bound form or as a free-soluble form. The soluble form activates both osteoclast differentiation and activation, thus initiating bone resorption [2]. A fine regulation of the cross-talk between these two cells is provided by the synthesis of another protein, osteoprotegerin (OPG), a soluble receptor for RANKL that is secreted by osteoblasts and bone marrow stromal cells to decoy the recognition of RANK by RANKL [3]. By binding RANKL, OPG prevents the ligand recognition of RANK, thus inhibiting osteoclast differentiation and activity and allowing new bone tissue deposition [4].

Indeed, the balanced sequence of interactions of the RANK/RANKL/OPG triad represents a tightly controlled system at the centre of the osteoclast-osteoblast functional unit that determines skeletal mass at any time in an organism's life. It is well established that the dysregulation of the RANK/RANKL/OPG triad significantly affects osteoclast maturation, survival, and activity, resulting in disorders of the bone remodelling cycle such as osteoporosis, Paget's disease, rheumatoid arthritis, and bone metastases [5,6]. Current pharmacological interventions are designed to either reduce bone resorption or stimulate bone formation. Drugs used to reduce bone resorption include hormone replacement therapy (HRT) and selective oestrogen receptor modulators (SERM), bisphosphonates and those used to increase bone formation include teriparatide (PTH 1-34), and strontium renalate [7,8]. However, these drugs have associated side effects such as hot flashes, osteonecrosis, and poor fracture healing and osteointegration [8]. Moreover, as RANKL plays other physiological roles including immune function, lymph node formation, and mammalian development [9], the development of therapeutic drugs that can replace its use with no side effects is desirable. One of the notable developments in this regard is the recently approved drug, Denosumab, a fully human monoclonal anti-RANK antibody that mimics the activities of OPG, which is used as a bone anti-resorptive agent [10,11]. However, this drug is expensive to produce and store, carries the risk of viral and prion contamination, and its effect on immune function is unclear [11]. For these reasons, the use of small molecules which are able to target remodeling pathways has been advocated [12]. To this end, peptides such as the gap-junction protein (connexin 43) mimetic (GAP27) [13], the RANK-mimetic peptide [14], the OPG-mimetic peptides [15], the TNF-[alpha] and the RANKL antagonist peptide [16], and a TNF receptor loop peptide mimic [17] have been identified as able to reduce bone resorption and/or promote bone formation.

This work reports the synthesis of a known bone specific cyclic OPG-mimetic peptide (OP3-4) sequence YCEIEFCYLIR, which is based on residues 113–122 of the human OPG sequence [15,18,19], and its complexation with a contrast agent for magnetic resonance imaging (MRI), gadolinium, through a chelating moiety, 1,4,7,10-tetraazacyclododecane-1,4,7,10-tetraacetic acid (DOTA) [20]. The novel peptide formed, DOTA-Gd-OP3-4, was tested for MRI traceability and its direct inhibitory effect on osteoclastogenesis in vitro to offer a novel theranostics agent with the potential to enable the simultaneous diagnosis (or follow up) and treatments of bone affected by metabolic diseases.

2. Materials and Methods

2.1. Materials

Amino acids protected by 9-fluorenylmethyloxy carbonyl (Fmoc) where purchased from NovaBiochem, Watford, UK. Tenta Gel S NH$_2$ resin and Rink amide linker were purchased from Iris Biotech, Marktredwitz, Germany. Dimethylformamide (DMF), dichloromethane, diethyl ether, and methanol were of analytical grade ($\geq$99.5%) and obtained from Fisher Scientific, Loughborough, UK. Triisopropylsilane (TIPS, $\geq$98.0% GC), phenol ($\geq$99%), trifluoroacetic acid (TFA, $\geq$99%), and *N*,*N*-diisopropylethylamine (DIPEA, $\geq$99%) were obtained from Iris Biotech GmbH. 2-(1*H*-benzotriazole-1-yl)-1,1,3,3-tetramethyluronium hexafluorophosphate (HBTU, $\geq$97%) was purchased from NovaBiochem, UK. Tri-*tert*-butyl 1,4,7,10-tetraazacyclododecane-1,4,7,10-tetraacetate (DOTA-tri-*t*-Bu-ester) was purchased from Sigma Aldrich, Dorset, UK. All other reagents were of analytical grade.

2.2. Synthesis of Peptides

The peptides OP3-4 and DOTA tethered OP3-4 (DOTA-OP3-4) were synthesised by solid-phase peptide synthesis (SPPS) using the conventional Fmoc protection/deprotection strategy on Tenta Gel

S NH$_2$ resin (0.1 mmol equivalent) with DMF as the reaction solvent. The resin was pre-swollen in 7 mL DMF in a 10 mL reaction vessel and the peptides synthesised by first attaching an acid-liable Fmoc-Rink Amide linker to the resin followed by sequential amino acid coupling and deprotection steps as per peptide sequence. HBTU and DIPEA were used for the coupling reaction (×2, 30 min, room temperature) at one and two times the concentration of the amino acids and 20% (*v*/*v*) piperidine in DMF was used for deprotection (2 min, ×3). In all preparations, the resin, linker, and amino acids were added in the molar ratio of 1:4:4, respectively. Each coupling or deprotection step was followed by washing steps (×3 with DMF). After the final deprotection step, the resin was washed with 10 mL DMF (×3), dichloromethane (×4), methanol (×4), and diethylether (×4). The peptides were then dried in a vacuum oven at room temperature until reaching a constant weight (~2 h) and stored at 4 °C for further studies.

In the synthesis of DOTA-OP3-4 (Figure 1), Fmoc-Lys (Mtt)-OH was first coupled to the Rink-amide linker. The Mtt (4-methyltrityl) protecting group was then removed by a series of washes (1 min, ×9) with 1% TFA (*v*/*v*) in DMF until clear; Mtt appears yellow in DMF solution. DOTA-tri-*t*-Bu-ester was then coupled to lysine, followed by the deprotection of the Fmoc protected terminal on lysine. Two glycine amino acids were then sequentially coupled as spacers, followed by the subsequent addition of amino acids as per the OP3-4 sequence (Figure 1).

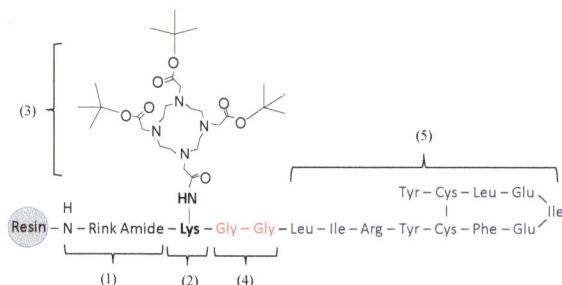

Figure 1. Schematic representation of the assembly of DOTA-OP3-4 by SPPS. Numbered sections indicate the order of assembly. The numbers used refer to: (**1**) the attachment of the rink amide linker; (**2**) the coupling of Fmoc lys (Mtt)–OH; (**3**) the removal of Mtt groups followed by the coupling of DOTA-Tri-t-Bu-ester; (**4**) the coupling of two glycines to provide a spacer; and (**5**) the coupling of the amino acids as per the OP3-4 peptide sequence and subsequent Cys-Cys disulphide bond cyclisation.

The peptides were freed from the resin by incubation in a cleavage solution (88% TFA, 5% H$_2$O, 5% phenol and 2% TIPS) for 3 h at room temperature. The cleaved peptides were filtered through glass wool, precipitated, and isolated by a series of washing and centrifugation steps in cold diethylether (≤4 °C). The samples were then dried over a stream of nitrogen gas and stored at −20 °C. Crude peptides were dissolved in 0.1% formic acid in methanol at 2 mg/L and characterised on an ion trap mass spectrometer (ITMS) model HTC Plus and a Time-of-Flight mass spectrometer (TOF-MS) model MicrOTOF, both from Bruker Daltonics (Billerica, MA, USA). The systems were optimised for the detection of the peptide *m/z* in every instance and once the peptide was characterised, the systems were re-optimised for the isolation and collection of the purified peptide. Data acquisition was carried out by Compass 1.1, Esquire 5.3, and Hystar 3.1 software (Bruker Daltonics, Billerica, MA, USA).

2.3. Cyclisation of OPG Mimetic Peptides

OP3-4 and DOTA-OP3-4 were cyclised by dimethyl sulfoxide (DMSO) oxidation to form cysteine-cysteine disulfide bonds as described in the literature [21], with slight modification. Briefly, peptides (1 mg) were dissolved in 60 mL of oxidising buffer (100 mM NaH$_2$PO$_4$ and 2 mM Gdn·HCl,

5% DMSO, pH 7.0) and shaken for 12 h. The solution was then acidified with 1 M HCOOH (250 μL) and the peptides purified and analysed by hydrophilic interaction liquid chromatography (HILIC) coupled to ITMS (LC-MS). Cyclisation (disulfide bridge formation) was confirmed by the quantitation of free thiol groups using Ellman's reagent following kit instructions (Invitrogen Molecular Probes T6060 Thiol and Sulfide Quantitation Kit, from ThermoFisher Scientific, Loughborough, UK) [22].

2.4. Gadolinium Chelation into DOTA-OP3-4

Cyclised DOTA-OP3-4 (1 mmoles, in 2 mL ethanol) was mixed with $GdCl_3 \cdot 6H_2O$ (1 mmol) in 5 mL water and the pH adjusted to 6.5 with Na_2CO_3 (0.1 mL) and stirred for 15 h at 60 °C. The peptides were then purified by high pressure liquid chromatography (HPLC) using a Solvent A (methanol, 0.1% TFA) and Solvent B (water, 0.1% TFA) gradient from 5% to 60% solvent. The system was coupled to the non-destructive soft-ionisation electrospray ITMS.

2.5. Chromatographic Separation of the Peptides by LC-MS

The peptide synthesis products were separated by HILIC on an HP 1200 HPLC (Agilent Technologies, Santa Clara, CA, USA) equipped with a TSK-Gel Amide 80 HILIC column (250 mm (length) × 4.6 mm internal diameter and 5 μm particle size) from TOSOH Bioscience (Tokyo, Japan). The system used a binary mobile phase of solvent A (acetonitrile, 0.1% (v/v) formic acid) and solvent B (water, 0.1% (v/v) formic acid) at a flow rate of 300 μL/min. The gradient elution program was: 0–50% solvent B for 0–20 min for OP3-4 and 5–90% solvent B for 0–45 min for both DOTA-OP3-4 and DOTA-Gd-OP3-4. The peptides were dissolved in 0.1% (v/v) formic acid in methanol at 20 μg/g and the injection volume was 100 μL. The tube from the column was divided into two channels of equal length and diameter, with one leading to the detector (ITMS) while the other led to the sample-collecting vessel where purified peptides were collected at the point of detection. The purified peptides were lyophilized and stored at −20 °C.

2.6. Tandem MS

ITMS was first optimised to detect the putative ions from the SPPS products in the full scan mode, followed by their isolation in the single ion monitoring mode (SIM). Energy was then applied to the system to fragment the isolated putative ion (MS/MS or MS2) and the resulting fragmentation products were scanned in product ion mode. Depending on the intensity and amount of the fragmentation products from MS/MS, the most intense fragment (base peak) was isolated and fragmented further (MS3), which made further confirmation of the identity of the peptide possible. The process was repeated with subsequent base peaks until the obtained peaks were within the noise signal.

2.7. MRI Analysis

The peptides (DOTA-OP3-4 and DOTA-Gd-OP3-4) were first dissolved in a minimum volume of DMSO and then diluted to a desired peptide concentration (0–10 μg/mL and 0.5% by volume DMSO) in phosphate-buffered saline (PBS) and the pH adjusted to 7.2 using 0.1 M HCl. DOTA-Gd and DOTA-OP3-4 were used at 10 μg/mL as the positive and negative controls, respectively. The DOTA-Gd-OP3-4 sample was then double diluted ten times from 10 μg/mL to 0.005 μg/mL and each sample (1 mL) was placed into different wells of a 24-well tissue culture plate for MRI analysis. The analysis was performed in T1 weighted scan mode using the Siemens AVANTO 1.5T MRI scanner with parameters set as: echo time (TE) 13, repetition time (TR) 200, and SL 1 mm. The analysis was conducted at the Clinical Imaging Science Centre, Brighton and Sussex Medical School, Brighton, UK.

2.8. Cell Viability Studies

Osteoblastic cells, the SaOS-2 cell line (ATCC, Manassas, VA, USA), were cultured to 85% confluence in McCoy's 5A culture medium (without L-glutamine) supplemented with 10% foetal bovine serum (PAA Laboratories GmbH, Cölbe, Germany). The cells were then trypsinised (5 mL trypsin, 37 °C for 6 min), washed, and seeded into 96-well plates at a 10^4 cells/well seeding density. The cells were then treated with different peptide concentrations (0–200 µM) and incubated (37 °C, 5% CO_2) for 24 h, after which the cell viability and cytotoxicity were evaluated.

Cell viability was studied by assessing the ability of the cells to metabolically reduce a tetrazolium salt 3-(4,5-dimethylthiazol-2-yl)-5-(3-carboxymethoxyphenyl)-2-(4-sulfophenyl)-2*H*-tetrazolium (MTS) to a soluble formazan derivative following kit instructions (CellTiter 96® AQueous One Solution Cell Proliferation Assay from Promega, Southampton, UK), with slight modification. Briefly, cells in a 96 well plate were washed three times with PBS and re-supplemented with 100 µL of culture medium per well. MTS/PMS (20 µL) solution (1:20 v/v) was then added to each well and the cells incubated for 2 h. Control cells were not exposed to peptides. The amount of formazan produced (purple colour) was measured at an absorbance wavelength of 490 nm using the Biochrom Asys UVM 340 plate reader equipped with Micro Win 2000 software (Biochrom Ltd., Cambridge, UK).

Cytotoxicity of the peptides was studied by assessing the amount of lactate dehydrogenase (LDH) released from cells as a measure of the number of lysed cells following kit instructions (CytoTox 96® Non-Radioactive Cytotoxicity Assay, Promega, UK). Briefly, after peptide treatment, cell culture medium isolated from every well was centrifuged (2000 rpm, 5 min) and the debris-free supernatant transferred to a clean 90 well plate. CytoTox 96® Reagent (100 µL) was added to each well and incubated for 30 min. The kit stop solution was added and absorbance measured at 490 nm. Positive control cells were completely lysed with Triton-X before centrifugation to estimate the maximum toxicity value in the same culture conditions measured as total LDH into the tissue culture supernatant. Negative control cells were not lysed. All control cells were not exposed to peptides.

Epi-fluorescence microscopy analysis for viable cells in cultures treated with 100 µM of each peptide was performed after Hoechst 33,258/Propidium iodide (H/PI, from Invitrogen, Carlsbad, CA, USA) nuclear staining (50 ng/mL) at an H:PI ratio of 1:1 per well in a 24 well culture plate. Diffusely H positive nuclei (blue and live) and diffusely PI positive nuclei (pink and dead) were counted and the number of live cells was presented as a percentage of the total number of cells per field. At least 200 cells from different randomly picked areas were counted per well.

2.9. Monocyte Isolation and Osteoclastogenesis Studies

Ethical approval for human blood studies was acquired from the University of Brighton Faculty Research Ethics and Governance Committee, Brighton, UK. Human whole blood (6 mL) from healthy consented volunteers was layered on Histopaque (3 mL, density: 1.077 g/mL, from Sigma Aldrich, UK) in heparinised tubes (9 mL Vacuette NH Sodium Heparin, Greiner Bio-One International GmbH) and centrifuged ($450\times g$, 30 min). The monocyte-rich buffy coat layer was isolated by gentle pipette aspiration and washed in 8 mL PBS ($\times3$) by centrifugation ($250\times g$, 15 min) to remove the platelet-rich plasma fraction. The cells were then resuspended in 1 mL αMEM (PAA Laboratories GmbH, Germany) medium and seeded into 24 well plates at a 5×10^5 cells/well seeding density. After a 3-h incubation (37 °C, 5% CO_2), non-adherent cells were rinsed with PBS ($\times3$) and the adherent cells cultured for 24 h in αMEM medium supplemented with 10% (v/v) FBS and Penicillin/Streptomycin ($1\times$).

Monocyte differentiation into osteoclasts was induced by spiking the cells with human recombinant RANKL (rh RANKL) and Macrophage Colony Stimulating Factor (rh M-CSF), both purchased from Invitrogen, UK. Firstly, monocytes were treated with different concentrations of rh RANKL (0, 10, 50, and 100 ng/mL) in the presence of 25 ng/mL rh M-CSF and over different time periods (two, four, and six days). Osteoclastogenesis inhibition was then studied by culturing monocytes in αMEM medium supplemented with 10% (v/v) FBS, Penicillin/Streptomycin ($1\times$), 50 ng/mL rh RANKL, and 25 ng/mL rh M-CSF for four days in the presence or absence of 50 µM peptides (OP3-4, DOTA-OP3-4 and DOTA-Gd-OP3-4). The positive control was treated with rh OPG (50 ng/mL) and the negative control had no treatment. The degree of osteoclastogenesis was assessed by tartrate-resistant acid phosphatase (TRAP) staining using a commercial kit (387A for cytochemical staining from Sigma Aldrich, UK) according to the manufacturer's instructions. TRAP positive multinucleated cells were counted using a Nikon Eclipse T*i*-U microscope equipped with a Nikon DIGITAL SIGHT DS-Fi1 camera (Nikon Corporation., Tokyo, Japan) and NIS-Elements BR 3.2 software (Nikon Corporation., Tokyo, Japan) was used to count the cells in six different randomly selected areas in each well.

3. Statistical Analysis

All statistical analyses were performed by using analysis of variance (ANOVA) (software: Minitab Version 15, Minitab Inc, State College, PA, USA). Whenever ANOVA indicated the groups were significantly different, a *t*-test for independent samples was performed. Samples were considered significantly different at $p \le 0.05$.

4. Results and Discussion

4.1. Peptide Synthesis and Characterisation

Due to the lack of peptide standards for comparative verification of the successful synthesis of the putative peptides, two mass spectrometry analysis approaches were used: (1) TOF-MS, to benefit from its high mass resolution and accuracy (up to five decimal places), high linearity, and wide scan range (50 to 3000 *m/z*); and (2) ITMS, to benefit from its high sensitivity and capacity to sequentially trap and fragment desired ions so as to identify the building units of a molecule that would confirm successful synthesis.

The MS acquisition for the putative peptides was first performed in full scan mode using TOF-MS, providing an overview of the products. The theoretical *m/z* of the peptides and their respective fragment ions were generated using the MS-product tool of the online program ProteinProspector V5.10.0. [23]. Successful cyclisation through the cysteine-cysteine disulfide bridge formation was confirmed by the absence (95% reduction) of free thiol groups in both OP3-4 and OP3-4-DOTA after the oxidation reaction (data not shown).

The molecular structure and mass spectra of the OP3-4 peptide (*m/z* 1450.7) synthesis products acquired by TOF-MS are shown in Figure 2A, where the putative ions $[M + 2H]^{2+}$ (*m/z* 725.87, base peak) and $[M + H]^{+}$ (*m/z* 1450.68) are annotated. Although incomplete fragments were seen in the spectra of OP3-4, the fragments were not sufficiently abundant to allow their conclusive identification from the background noise. The analysis of the synthesis products of the novel DOTA-OP3-4 peptide by TOF-MS showed numerous undesired product ions. As a result, ITMS, with its ability to trap specific desired ions using the single ion mode (ITMS-SIM), was used to confirm the presence of the putative ion. The full scan MS spectrum of the synthesis products of the novel DOTA-OP3-4 peptide (*m/z* 2080.2) showed ions with *m/z* 694.1 (base ion) and *m/z* 1040.5, which correspond to the putative peptide ions $[M + 3H]^{3+}$ and $[M + 2H]^{2+}$, respectively (Figure 2B). The introduction of a lysine-glycine-glycine spacer between the OP3-4 sequence and the DOTA moiety in DOTA-OP3-4 was essential to avoiding potential steric hindrance during the synthesis and to maximise the exposure of the peptide to the relative cell receptor. The numerous product ions observed in the MS spectra of DOTA-OP3-4 may be the effect of using 0.1% (*v/v*) TFA in the presence of the cysteine-rich peptide during the additional step to remove lysine protecting groups (Mtt) and the subsequent coupling of DOTA. The interaction of both the terminal carboxyl groups on the growing peptides and of the DOTA moiety with the side chain groups on the peptide during the synthesis and the cleavage steps may also have led to intra- and intermolecular interactions, resulting in numerous incomplete and undesired peptidic chains.

The analysis of the OP3-4 peptide by LC-MS showed the peptide to account for 80% of the crude synthesis products and a purity of >98% could be achieved after LC-MS purification with ITMS as the detector (Figure 3A). Despite the presence of numerous intense peaks corresponding to undesired ions, the ion corresponding to DOTA-OP3-4 was found to account for 60% of the crude synthesis products by LC-MS. This disparity in the intensity of the synthesis products by TOF-MS and LC-MS coupled to ITMS could be due to differences in the volatility of the ions in the MS [24] as smaller molecules are more volatile than larger ones in MS. Although TOF-MS gives highly accurate *m/z* for the ions, the system is not as sensitive as the ITMS. Since ITMS allows for optimisation of the detection of the desired ions within a specified *m/z* range, and LC-MS separates molecules according to their polarity, a more accurate result of the purity and/or yield of the synthesis could be obtained. Indeed, after isolation of the putative ion, DOTA-OP3-4 was found to account for >95% of the isolated product (Figure 3B).

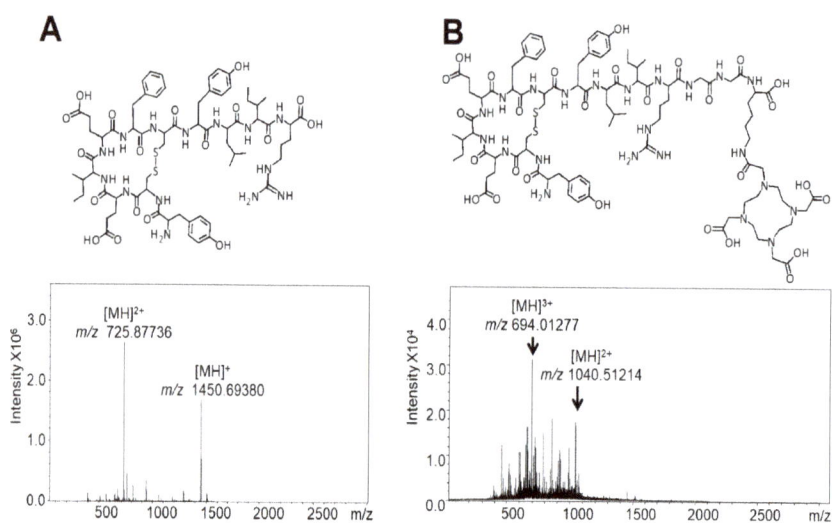

Figure 2. TOF-MS spectra and Structure of the peptides: (**A**) OP3-4 and (**B**) DOTA-OP3-4.

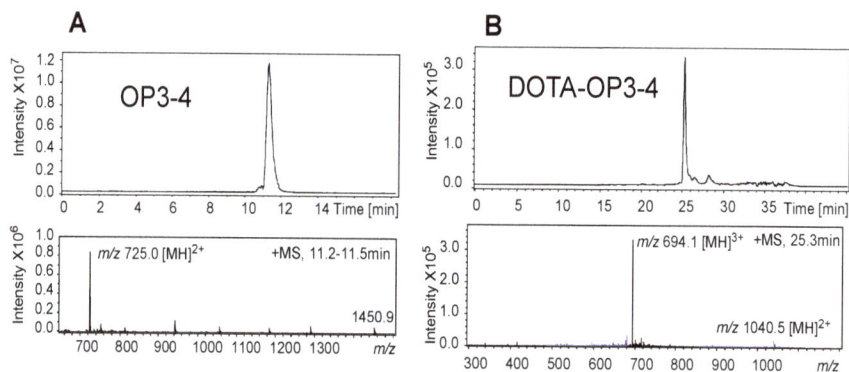

Figure 3. LC-MS chromatographic separation profile of the putative peptides (top) (**A**) OP3-4 (*m/z* 1450.9) and (**B**) DOTA-OP3-4, (*m/z* 2081.4) with their respective ITMS ion mass spectra (bottom) eluted as peaks at 11.2–11.5 and at 25.3 min. The purity of the putative peptides, estimated by integration of the peak areas, was >95% (OP3-4) and 90% (DOTA-OP3-4).

Successful chelation of the gadolinium ion (Gd^{3+}) onto purified DOTA-OP3-4 to produce DOTA-Gd-OP3-4 (Figure 4A) was confirmed by LC-MS, where the peak with *m/z* 694.1 for DOTA-OP3-4 peptide shown in Figure 3B was replaced by a base peak with *m/z* 714.8, corresponding to the ion [M + 3H]$^{3+}$ of DOTA-Gd-OP3-4 after the chelation of Gd^{3+} (Figure 4B). LC-MS allowed the isolation of the peptide from salts, non-chelated Gd^{3+}, and other synthesis products, and the analysis of the purified DOTA-Gd-OP3-4 sample showed no evidence of residual free Gd^{3+} ions.

Gadolinium is a lanthanide metal commonly used in clinics as an MRI contrast agent. Gadolinium is used in its ionic form chelated by large, non-toxic linear or macrocyclic organic molecules that form stable and biochemically inert gadolinium contrast agents (GCA). This sequestration of Gd^{3+} with organic chelates circumvents the known toxicity of free Gd^{3+} in humans and ensures their excretions from the body through kidney clearance. However, these GCA, in particular those of linear chelates such as Omniscan (gadodiamide), Magnevist (gadopentetate dimeglumine), and Optimark (gadoversetamide), are associated with slow excretion rates due to accumulation within the body (i.e., in the liver) and toxicity from free Gd^{3+} ion release due to transmetallisation with Zn^{2+}, Cu^{2+}, and Ca^{2+} in vivo [25,26]. Macrocyclic GCAs such as dotarem (Guerbet) and gadovist (Bayer Schering Pharma AG) are less toxic as they bind Gd^{3+} more tightly and are more stable with low dissociation rates than linear chelates. However, these GCAs, like linear ones, are associated with short circulation in the body and inefficient discrimination between diseased and normal tissue seen in use of low molecular diethylenetriaminepentaacetic (DTPAs) [25,26].

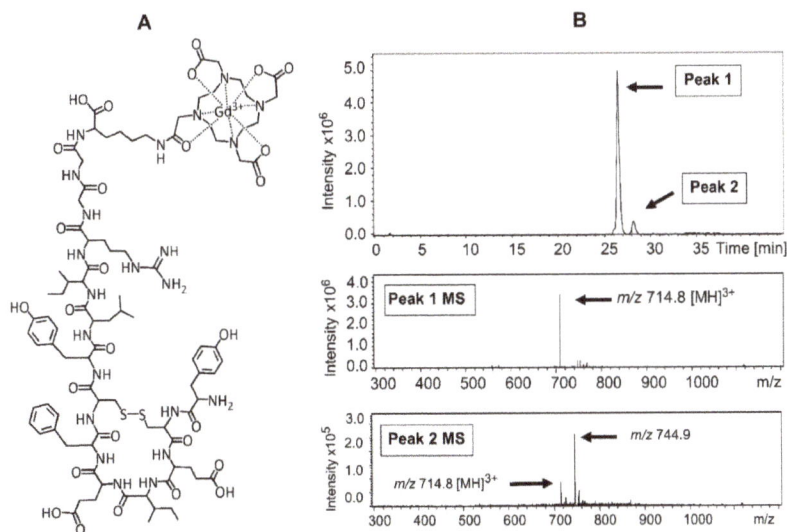

Figure 4. (**A**) Schematic representation of DOTA-OP3-4 with Gd^{3+} conjugate (DOTA-Gd-OP3-4); (**B**) Chromatographic separation profile of the purified DOTA-Gd-OP3-4. The ITMS spectra of the ions eluted in peak 1 (putative peptide) and peak 2 (contaminating fraction) are also shown (bottom). The purity of the peptide, estimated by integration of the peak areas, was >90%.

Through this work, tissue-specific imaging may be improved by the incorporation of a peptide sequence specific for the cross talk of bone cells, the RANK-RANKL pathway, and the potential toxicity of Gd^{3+} is circumvented by the use of DOTA, a macrocyclic chelate for Gd^{3+} [27]. Indeed, some DOTA-based Gd contrast agents are FDA approved and are among the most widely used contrast agents in clinical imaging with well-established safety and MRI efficacy [27,28].

In the chelation of Gd^{3+} into DOTA-OP3-4, the DOTA moiety acts as a polydentate ligand and envelops the metal cations, in this case complexing Gd^{3+}, to give an MRI visible peptide.

The coordination of the DOTA ligands and metal ion in the complex depends on the conformation of the ligand and geometric tendencies of the metal cation [29,30]. On its own, DOTA acts as an octadentate ligand, binding the metal through four amino and four carboxylate groups. In this study, the DOTA molecule acts as a septadentate since one of the carboxylate groups is used in the formation of a covalent bond with the peptide. However, a carboxylate group from the amino acid linking DOTA and the peptide provides the eighth ligand and restores the octadentate state, forming a highly stable coordination complex [30].

Analysis of the result obtained from a T1 weighted MRI scan confirmed the successful chelation of Gd^{3+}, with DOTA-Gd-OP3-4 producing a brighter signal than DOTA-OP3-4, which had a signal similar to the sample with PBS solutions only (Figure 5). Moreover, the intensity of the T1 signal from DOTA-Gd-OP3-4 samples was directly proportional to the concentration of the peptide.

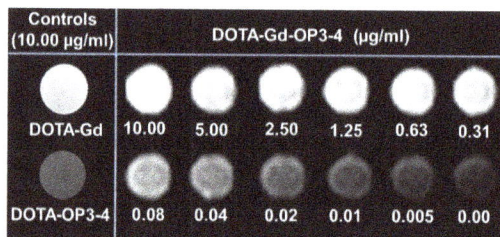

Figure 5. T1 weighted MRI scan of DOTA-Gd-OP3-4 showing a concentration-dependent signal intensity. The peptides, pre-dissolved in DMSO, were diluted to the desired concentration in PBS buffer (1 mL) and the MRI signal acquired at the same time using the parameters: SL5, TE 8.7, and TR550. DOTA-Gd (MW 557.64 g/mol) and DOTA-Gd-OP3-4 (MW 2141.4 g/mol) were prepared at a concentration of 1 µg/ml, corresponding to 1.79 µM and 0.47 µM Gd, respectively.

The coordination of the DOTA ligands and metal ion in the complex depends on the conformation of the ligand and geometric tendencies of the metal cation [29,30]. On its own, DOTA acts as an octadentate ligand, binding the metal through four amino and four carboxylate groups. In this study, the DOTA molecule acts as a septadentate since one of the carboxylate groups is used in the formation of a covalent bond with the peptide. However, a carboxylate group from the amino acid linking DOTA and the peptide provides the eighth ligand and restores the octadentate state, forming a highly stable coordination complex [30].

Although the potential theranostic benefits of the developed peptide are clear, the minimum peptide concentration that would be required for a visible change in the MRI signal intensity in vivo remains to be determined through future relaxivity studies [31,32]. Indeed, relaxivity calculation requires the different concentrations of the peptide to be analysed separately in T1 and T2 modes and using the respective r1 and r2 to determine relaxivity [31–33]. Although the in vivo relaxivity and safety of the DOTA-Gd moiety are well-established, [28,32,34], the new peptide complex presents a different immediate molecular environment that can affect the MRI profile of Gd^{3+} [35,36]. The importance of relaxivity studies in vivo is supported by the effect that the molecular environment has on the MRI profile, leading investigators to question the clinical efficacy of T1 relaxivity measurements performed in vitro in simple media and temperatures different from those of living tissue [32–34,36].

In addition, although TOF-MS and LC-MS demonstrated the successful synthesis of peptides, it was necessary to perform tandem MS for additional confirmation of the results due to the lack of standards. The ions to be fragmented were acquired in product ion scan mode, isolated, and further fragmented to produce smaller units that allowed peptide identification and the confirmation of successful synthesis. In all cases, the voltages for the optimal transfer of the ions from the LC into the ion trap, their stabilisation in the trap, and fragmentation were optimised. A list of the peptides, isolated ions thereof (SIM), and fragmentation products are given in Table 1.

Table 1. Tandem MS of OP3-4 and DOTA-OP3-4 peptides using ITMS.

	TOF-MS (Full Scan MS)			ITMS MS-MS Product Ions (Ion Trap)		
Peptide	(*m/z*)	Tentative Assignment	SIM (*m/z*)	MS/MS Product Ion (*m/z*)	Tentative Assignment	MS/MS Scan Range
OP3-4	1451.6960	[M + H]$^+$				500–1500
	726.377808	[M + 2H]$^{2+}$	726	886	b$_7$-2H	300–1200
				733	b$_6$-H$_2$O	
				715	b$_6$-2H$_2$O	
				564	y$_4$	
DOTA-OP3-4	1040.52070	[M + 2H]$^{2+}$		987		300–1200
	694.01938	[M + 3H]$^{3+}$	694	691	b$_{11}$-2H$_2$O-NH$_3$$^{2+}$	
				596		
				514	y$_7$$^{2+}$	
				458	y$_6$$^{2+}$	
				401	y$_5$$^{2+}$	

4.2. The Effect of the Peptides on Cell Viability and Osteoclastogenesis

The effects that the peptides have on cells was first studied by spiking SAOS-2 cells with different concentrations of OP3-4. MTS and LDH assay results showed cell viability within a wide range of peptide concentrations (Figure 6A). The H/PI study of the effect of OP3-4-DOTA and OP3-4-DOTA-Gd on the SAOS-2 cells relative to OP3-4 at a fixed concentration (100 µM) showed no significant effect on cell viability (Figure 6B). Although there appears to be a decrease in cell viability from the control to OP-3-4 to DOTA-OP3-4 and DOTA-Gd-OP3-4, the differences were not significant to the control and across the different treatment groups ($p > 0.05$).

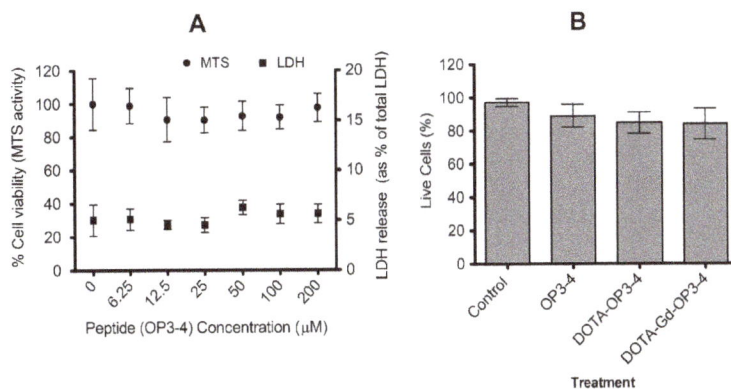

Figure 6. OP3-4, DOTA-OP3-4, and DOTA-Gd-OP3-4 have no significant effect on the viability of SAOS-2 cells. (**A**) The cells were cultured with different concentrations of OP3-4 peptide for 24 h and the percentage of dead and live cells determined by assaying for LDH in supernatant and MTS assay, respectively. Absorbance was measured at 490 nm. (**B**) The effect of derivatised OP3-4 peptides (at 100 µM) on cell viability was determined by fluorescence microscopy after H/PI live and dead cell staining (**B**). Data is mean ± SE, *n* = 6.

4.3. Osteoclastogenesis Inhibition Studies

The studies conducted to assess the effect of cytokine (rh RANKL and rh M-CSF) concentration over two, four, and six days in culture showed that the amount of TRAP-positive giant multinucleated cells (MNC) per well increased with the increase of both cytokine (rh RANKL) concentration and culturing time (Figure 7A). The number of TRAP-positive MNCs increased significantly after four days in cultures treated with 50 ng/mL and 100 ng/mL rh RANKL. However, osteoclastogenesis in cultures treated with 50 ng/mL rh RANKL was not significantly different from 100 ng/mL at day 4 and 6 ($p > 0.05$). As such, the rh RANKL concentration of 50 ng/mL and a culture time-period of four days were adopted for studies on osteoclastogenesis inhibition by the peptides.

Representative images of the cells after four days of culture in the presence or absence of the cytokines and in the presence or absence of the peptides are given in Figure 7B. The cells that did not receive cytokines and those treated with rh RANKL only were TRAP negative, morphologically small, and round, with a tendency to form clusters (Figure 7B(i and ii)). However, the fact that this clustering phenomenon was observed in cells that did not receive either cytokine suggests that it could be due to other factors apart from rh M-CSF and/or RANKL and more studies are necessary to clarify that. On the other hand, the cells treated with only rh M-CSF were TRAP negative and morphologically narrow and stretched (Figure 7B(iii)), while those treated with both cytokines differentiated into cells that were much larger ($> \times 10$), multinucleated, morphologically irregular, and TRAP positive (Figure 7B(iv)). This is consistent with the established knowledge that M-CSF and RANKL induce the migration of monocytes towards each other, resulting in cell fusion to produce a giant osteoclast like TRAP-positive MNC [37–39]. The size and amount of TRAP-positive MNCs was seen to reduce in cultures treated with the peptides (Figure 7B(v–vii)) and in the positive control (rh OPG) (Figure 7B(viii)). This result was also quantitatively confirmed by counting the number of TRAP-positive cells with each respective treatment (Figure 7C). The observed potency of OP3-4 in osteoclastogenesis inhibition is consistent with previous findings [17,18] and was significantly higher than that observed for DOTA-OP3-4 and DOTA-Gd-OP3-4. However, all the peptides significantly reduced the RANKL/M-CSF-induced osteoclastogenesis.

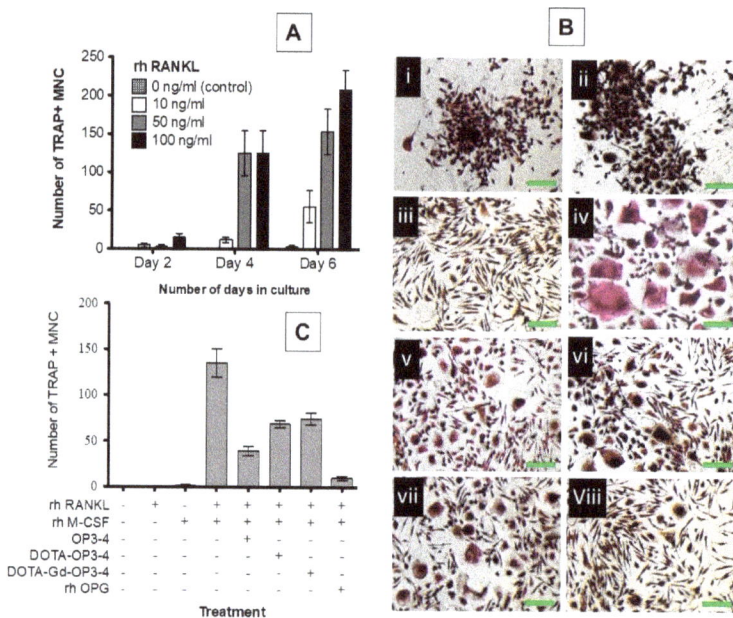

Figure 7. (**A**) Graphical representation of the effect of rh RANKL concentration and time in culture on osteoclastogenesis. (**B**) The effect of peptides on osteoclastogenesis after a four-day culture with (**i**) no cytokines or peptides, (**ii**) rh RANKL only, (**iii**) rh M-CSF only, (**iv**) rh RANKL and rh M-CSF, (**v**) OP3-4, (**vi**) DOTA-OP3-4, (**vii**) DOTA-Gd-OP3-4, and (**viii**) rh OPG (positive control): iv–ix had rh RANKL (50 ng/mL) and rh M-CSF (25 ng/mL). (**C**) Graphical representation of osteoclastogenesis inhibition by OPG mimetic peptides.

Nanomaterials **2018**, *8*, 399

Indeed, the presentation and binding of the developed OPG mimetic peptides to RANK may have been specific and sufficient to inhibit RANK-RANKL interaction. It is well established that although the molecules of the TNF superfamily (TNFSF), to which RANK and RANKL belong, are similar in structure, several members of this cytokine family show significant sequence diversity [40]. Consistent with this observation, studies on the 3D structures and mechanism of members of the TNFSF such as TNFα [41], TNF-β [42], RANKL/RANK [43,44], sTNF-R1[45], CD40L [46], TRAIL-DR5 [47], and many more, have revealed that these cytokines recognise their ligands or receptor with specificity and in most cases, exclusively [40,44]. In the present study, the potency of the rh OPG was significantly higher than that of the peptides although this could be, in part, due to the fact that OPG interacts with RANKL through multiple active sights and domains when compared to the peptides. However, the observed potency of the OPG mimetic peptides, which are smaller molecules compared to the OPG ligand, may indicate a higher degree of specificity that is enough for a desired effect of reducing osteoclastogenesis. It is also worth noting that the dosage used for peptides and OPG was not equimolar. Therefore, the data were not directly comparable due to the differences in the number of active domains and molecular weight of the natural protein and of the synthetic peptides. As such, the cultures treated with OPG should only be considered as a positive control.

The peptide is an OPG mimic designed to bind to RANKL and attenuate osteoclastogenesis. This in vitro cell work demonstrated the theranostics potential of the peptide, DOTA-Gd-OP3-4, in future clinical applications. However, in vivo evaluations of the peptide biodistribution, tissue retention time, and bone targeting properties are necessary to inform on the safety and potential clinical efficacy of the developed theranostics.

5. Conclusions

OPG mimetic peptide (OP3-4) and its novel derivatives DOTA-OP3-4 and MRI detectable DOTA -Gd-OP3-4 were successfully synthesised with DOTA-Gd-OP3-4, showing a concentration-dependent MRI T1 signal. The OP3-4 was found to be noncytotoxic on osteoblasts at the concentrations examined and its derivatisation into DOTA-OP3-4 and DOTA-Gd-OP3-4 did not reduce cell viability. Moreover, their potency in the inhibition of RANKL/M-CSF-induced osteoclastogenesis in monocytes was clearly proven. Therefore, this work provides evidence in support of a novel theranostics tool for a more efficient and specific treatment of metabolic bone diseases, such as osteoporosis and metastases.

Author Contributions: Conceptualization, L.M. and M.S.; Methodology, L.M. and M.S.; Validation, L.M., M.S., S.T.M., and R.B.; Formal Analysis, L.M. and M.S.; Investigation, L.M., R.B., N.G.D., and M.C.; Resources, M.S.; Data Curation, L.M., S.T.M., and M.S.; Writing—Original Draft Preparation, L.M.; Writing—Review & Editing, L.M., S.T.M., and M.S.; Visualization, L.M. and M.S.; Supervision, M.S. and S.T.M.; Project Administration, M.S.; Funding Acquisition, M.S.

Funding: This work was funded by the Orthopaedic Research UK, Furlong House, 10a Chandos Street, London W1G 9DQ, United Kingdom, www.oruk.org.

Acknowledgments: The authors wish to thank Dr Howard Dodd of the University of Brighton, UK, for his technical assistance with mass spectrometry.

References

1. Li, J.; Xiang, L.; Jiang, X.; Teng, B.; Sun, Y.; Chen, G.; Chen, J.; Zhang, J.V.; Ren, P.-G. Investigation of bioeffects of G protein-coupled receptor 1 on bone turnover in male mice. *J. Orthop. Transl.* **2017**, *10*, 42–51. [CrossRef] [PubMed]

2. Henriksen, K.; Bollerslev, J.; Everts, V.; Karsdal, M.A. Osteoclast Activity and Subtypes as a Function of Physiology and Pathology—Implications for Future Treatments of Osteoporosis. *Endocr. Rev.* **2011**, *32*, 31–63. [CrossRef] [PubMed]

3. Crockett, J.C.; Helfrich, M.H. Chapter 6—Early Onset Pagetic Disorders: Implications for Late-Onset Paget's Disease of Bone. In *Advances in Pathobiology and Management of Paget's Disease of Bone*; Reddy, S.V., Ed.; Academic Press: Cambridge, MA, USA, 2016; pp. 71–88. ISBN 978-0-12-805083-5.

4. Sigl, V.; Penninger, J.M. Chapter 8—RANK and RANKL of Bones, T Cells, and the Mammary Glands A2—Lorenzo, Joseph. In *Osteoimmunology (Second Edition)*; Horowitz, M.C., Choi, Y., Takayanagi, H., Schett, G., Eds.; Academic Press: San Diego, CA, USA, 2016; pp. 121–142. ISBN 978-0-12-800571-2.

5. Kohli, S.S.; Kohli, V.S. Role of RANKL–RANK/osteoprotegerin molecular complex in bone remodeling and its immunopathologic implications. *Indian J. Endocrinol. Metab.* **2011**, *15*, 175–181. [CrossRef] [PubMed]

6. Kearns, A.E.; Khosla, S.; Kostenuik, P.J. Receptor Activator of Nuclear Factor κB Ligand and Osteoprotegerin Regulation of Bone Remodeling in Health and Disease. *Endocr. Rev.* **2008**, *29*, 155–192. [CrossRef] [PubMed]

7. Appelman-Dijkstra, N.M.; Papapoulos, S.E. Modulating Bone Resorption and Bone Formation in Opposite Directions in the Treatment of Postmenopausal Osteoporosis. *Drugs* **2015**, *75*, 1049–1058. [CrossRef] [PubMed]

8. De Villiers, T. Johannes, Bone health and osteoporosis in postmenopausal women. *Best Pract. Res. Clin. Obstet. Gynaecol.* **2009**, *23*, 73–85. [CrossRef] [PubMed]

9. Weitzmann, M.N. The Role of Inflammatory Cytokines, the RANKL/OPG Axis, and the Immunoskeletal Interface in Physiological Bone Turnover and Osteoporosis. *Scientifica* **2013**, *2013*, 29. [CrossRef] [PubMed]

10. Rizzoli, R.; Yasothan, U.; Kirkpatrick, P. Fresh from the pipeline: Denosumab. *Nat. Rev. Drug Discov.* **2010**, *9*, 591–592. [CrossRef] [PubMed]

11. Anastasilakis, A.D.; Toulis, K.A.; Polyzos, S.A.; Anastasilakis, C.D.; Makras, P. Long-term treatment of osteoporosis: Safety and efficacy appraisal of denosumab. *Ther. Clin. Risk Manag.* **2012**, *8*, 295–306. [CrossRef] [PubMed]

12. Wang, D.; Miller, S.C.; Kopecková, P.; Kopecek, J. Bone-targeting macromolecular therapeutics. *Adv. Drug Deliv. Rev.* **2005**, *57*, 1049–1076. [CrossRef] [PubMed]

13. Ilvesaro, J.; Tavi, P.; Tuukkanen, J. Connexin-mimetic peptide Gap 27 decreases osteoclastic activity. *BMC Musculoskelet. Disord.* **2001**, *2*, 10. [CrossRef]

14. Hur, J.; Ghosh, A.; Kim, K.; Ta, H.M.; Kim, H.; Kim, N.; Hwang, H.-Y.; Kim, K.K. Design of a RANK-Mimetic Peptide Inhibitor of Osteoclastogenesis with Enhanced RANKL-Binding Affinity. *Mol. Cells* **2016**, *39*, 316–321. [PubMed]

15. Shin, J.; Kim, Y.-M.; Li, S.-Z.; Lim, S.-K.; Lee, W. Structure-Function of the TNF Receptor-like Cysteine-rich Domain of Osteoprotegerin. *Mol. Cells* **2008**, *25*, 352–357. [PubMed]

16. Alles, N.; Soysa, N.S.; Hussain, M.D.A.; Tomomatsu, N.; Saito, H.; Baron, R.; Morimoto, N.; Aoki, K.; Akiyoshi, K.; Ohya, K. Polysaccharide nanogel delivery of a TNF-[alpha] and RANKL antagonist peptide allows systemic prevention of bone loss. *Eur. J. Pharm. Sci.* **2009**, *37*, 83–88. [CrossRef] [PubMed]

17. Aoki, K.; Saito, H.; Itzstein, C.; Ishiguro, M.; Shibata, T.; Blanque, R.; Mian, A.H.; Takahashi, M.; Suzuki, Y.; Yoshimatsu, M.; et al. A TNF receptor loop peptide mimic blocks RANK ligand-induced signaling, bone resorption, and bone loss. *J. Clin. Investig.* **2006**, *116*, 1525–1534. [CrossRef] [PubMed]

18. Cheng, X.; Kinosaki, M.; Takami, M.; Choi, Y.; Zhang, H.; Murali, R. Disabling of Receptor Activator of Nuclear Factor- KappaB (RANK) Receptor Complex by Novel Osteoprotegerin-like Peptidomimetics Restores Bone Loss in vivo. *J. Biol. Chem.* **2004**, *279*, 8269–8277. [CrossRef] [PubMed]

19. Ta, H.M.; Nguyen, G.; Thi, T.; Jin, H.M.; Choi, J.; Park, H.; Kim, N.; Hwang, H.-Y.; Kim, K.K. Structure-Based Development of a Receptor Activator of Nuclear Factor-kB Ligand (RANKL) Inhibitor Peptide and Molecular Basis for Osteoporosis. *Proc. Natl. Acad. Sci. USA* **2010**, *107*, 20281–20286. [CrossRef] [PubMed]

20. Mbundi, L.; Meikle, S.T.; Santin, M. Biospecific Agents for Bone. WO2014122431, 31 January 2014.

21. Góngora-Benítez, M.; Tulla-Puche, J.; Paradís-Bas, M.; Werbitzky, O.; Giraud, M.; Albericio, F. Optimized Fmoc solid-phase synthesis of the cysteine-rich peptide linaclotide. *Pept. Sci.* **2011**, *96*, 69–80. [CrossRef] [PubMed]

22. Winther, J.R.; Thorpe, C. Quantification of Thiols and Disulfides. *Biochim. Biophys. Acta* **2014**, *1840*, 838–846. [CrossRef] [PubMed]

23. UCSF Mass Spectrometry Facility ProteinProspector v 5.10.0. Proteomics Tools for Mining Sequence Databases in Conjunction with Mass Spectrometry Experiments. Available online: http://prospector.ucsf.edu/prospector/mshome.htm (accessed on 21 October 2014).

24. Annesley, T.M. Ion Suppression in Mass Spectrometry. *Clin. Chem.* **2003**, *49*, 1041–1044. [CrossRef] [PubMed]

25. Perazella, M.A. Current Status of Gadolinium Toxicity in Patients with Kidney Disease. *Clin. J. Am. Soc. Nephrol.* **2009**, *4*, 461–469. [CrossRef] [PubMed]

26. Svenson, S.; Tomalia, D.A. Dendrimers in biomedical applications-reflections on the field. *Adv. Drug Deliv. Rev.* **2005**, *57*, 2106–2129. [CrossRef] [PubMed]

27. Rogosnitzky, M.; Branch, S. Gadolinium-based contrast agent toxicity: A review of known and proposed mechanisms. *Biometals* **2016**, *29*, 365–376. [CrossRef] [PubMed]

28. Zhou, Z.; Lu, Z.-R. Gadolinium-Based Contrast Agents for MR Cancer Imaging. *Wiley Interdiscip. Rev. Nanomed. Nanobiotechnol.* **2013**, *5*, 1–18. [CrossRef] [PubMed]

29. Stetter, H.; Frank, W. Complex Formation with Tetraazacycloalkane-N,N',N'',N'''-tetraacetic Acids as a Function of Ring Size. *Angew. Chem. Int. Ed. Engl.* **1976**, *15*, 686. [CrossRef]

30. Viola-Villegas, N.; Doyle, R.P. The coordination chemistry of 1,4,7,10-tetraazacyclododecane-N,N',N'',N''' -tetraacetic acid (H4DOTA): Structural overview and analyses on structure–stability relationships. *Coord. Chem. Rev.* **2009**, *253*, 1906–1925. [CrossRef]

31. Hathout, G.; Jamshidi, N. Parameter Optimization for Quantitative Signal-Concentration Mapping Using Spoiled Gradient Echo MRI. *Radiol. Res. Pract.* **2012**, *2012*, 815729. [CrossRef] [PubMed]

32. Overoye-Chan, K.; Koerner, S.; Looby, R.J.; Kolodziej, A.F.; Zech, S.G.; Deng, Q.; Chasse, J.M.; McMurry, T.J.; Caravan, P. EP-2104R: A Fibrin-Specific Gadolinium-Based MRI Contrast Agent for Detection of Thrombus. *J. Am. Chem. Soc.* **2008**, *130*, 6025–6039. [CrossRef] [PubMed]

33. Varga-Szemes, A.; Kiss, P.; Rab, A.; Suranyi, P.; Lenkey, Z.; Simor, T.; Bryant, R.G.; Elgavish, G.A. In vitro Longitudinal Relaxivity Profile of Gd(ABE-DTTA), an Investigational Magnetic Resonance Imaging Contrast Agent. *PLoS ONE* **2016**, *11*, e0149260. [CrossRef] [PubMed]

34. Shen, Y.; Goerner, F.L.; Snyder, C.; Morelli, J.N.; Hao, D.; Hu, D.; Li, X.; Runge, V.M. T1 Relaxivities of Gadolinium-Based Magnetic Resonance Contrast Agents in Human Whole Blood at 1.5, 3, and 7 T. *Investig. Radiol.* **2015**, *50*, 330–338. [CrossRef] [PubMed]

35. Bagher-Ebadian, H.; Paudyal, R.; Nagaraja, T.N.; Croxen, R.L.; Fenstermacher, J.D.; Ewing, J.R. MRI estimation of gadolinium and albumin effects on water proton. *NeuroImage* **2011**, *54*, S176–S179. [CrossRef] [PubMed]

36. Noordin, S.; Winalski, C.S.; Shortkroff, S.; Mulkern, R.V. Factors affecting paramagnetic contrast enhancement in synovial fluid: Effects of electrolytes, protein concentrations, and temperature on water proton relaxivities from Mn ions and Gd chelated contrast agents. *Osteoarthr. Cartil.* **2010**, *18*, 964–970. [CrossRef] [PubMed]

37. Jansen, I.; Vermeer, J.; Bloemen, V.; Stap, J.; Everts, V. Osteoclast Fusion and Fission. *Calcif. Tissue Int.* **2012**, *90*, 515–522. [CrossRef] [PubMed]

38. Susa, M.; Luong-Nguyen, N.-H.; Cappellen, D.; Zamurovic, N.; Gamse, R. Human primary osteoclasts: In vitro generation and applications as pharmacological and clinical assay. *J. Transl. Med.* **2004**, *2*, 1–12. [CrossRef] [PubMed]

39. Zambonin, Z.A.; Teti, A.; Primavera, M.V. Monocytes from circulating blood fuse in vitro with purified osteoclasts in primary. *J. Cell Sci.* **1984**, *66*, 335–342.

40. Naismith, J.H.; Sprang, S.R. Modularity in the TNF-receptor family. *Trends Biochem. Sci.* **1998**, *23*, 74–79. [CrossRef]

41. Jones, E.Y.; Stuart, D.I.; Walker, N.P.C. Structure of tumour necrosis factor. *Nature* **1989**, *338*, 225–228. [CrossRef] [PubMed]

42. Banner, D.W.; D'Arcy, A.; Janes, W.; Gentz, R.; Schoenfeld, H.-J.; Broger, C.; Loetscher, H.; Lesslauer, W. Crystal structure of the soluble human 55 kd TNF receptor-human TNF[beta] complex: Implications for TNF receptor activation. *Cell* **1993**, *73*, 431–445. [CrossRef]

43. Ito, S.; Wakabayashi, K.; Ubukata, O.; Hayashi, S.; Okada, F.; Hata, T. Crystal Structure of the Extracellular Domain of Mouse RANK Ligand at 2.2-Å Resolution. *J. Biol. Chem.* **2002**, *277*, 6631–6636. [CrossRef] [PubMed]

44. Lam, J.; Nelson, C.A.; Ross, F.P.; Teitelbaum, S.L.; Fremont, D.H. Crystal structure of the TRANCE/RANKL cytokine reveals determinants of receptor-ligand specificity. *J. Clin. Investig.* **2001**, *108*, 971–979. [CrossRef] [PubMed]

45. Naismith, J.H.; Devine, T.Q.; Kohno, T.; Sprang, S.R. Structures of the extracellular domain of the type I tumor necrosis factor receptor. *Structure* **1996**, *4*, 1251–1262. [CrossRef]

46. Karpusas, M.; Hsu, Y.-M.; Wang, J.-H.; Thompson, J.; Lederman, S.; Chess, L.; Thomas, D. 2 å crystal structure of an extracellular fragment of human CD40 ligand. *Structure* **1995**, *3*, 1031–1039. [CrossRef]

47. Mongkolsapaya, J.; Grimes, J.M.; Chen, N.; Xu, X.-N.; Stuart, D.I.; Jones, E.Y.; Screaton, G.R. Structure of the TRAIL-DR5 complex reveals mechanisms conferring specificity in apoptotic initiation. *Nat. Struct. Mol. Biol.* **1999**, *6*, 1048–1053.

MDPI

St. Alban-Anlage 66

4052 Basel

Switzerland

Tel. +41 61 683 77 34

Fax +41 61 302 89 18

www.mdpi.com

Nanomaterials Editorial Office

E-mail: nanomaterials@mdpi.com

www.mdpi.com/journal/nanomaterials